ILLINOIS CENTRAL COLLEGE 3

W9-BQJ-222

WITHDRAWN

P 106 .R586 1989

Rorty, Richard.

Contingency, irony, and
  solidarity   79820

Illinois Central College
Learning Resources Center

Contingency, irony, and solidarity

RICHARD RORTY
University Professor of Humanities,
University of Virginia

I.C.C. LIBRARY

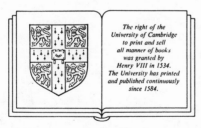

The right of the
University of Cambridge
to print and sell
all manner of books
was granted by
Henry VIII in 1534.
The University has printed
and published continuously
since 1584.

CAMBRIDGE UNIVERSITY PRESS
CAMBRIDGE
NEW YORK   PORT CHESTER   MELBOURNE   SYDNEY

79820

P
106
.R586
1989

Published by the Press Syndicate of the University of Cambridge
The Pitt Building, Trumpington Street, Cambridge CB2 1RP
40 West 20th Street, New York, NY 10011, USA
10 Stamford Road, Oakleigh, Melbourne 3166, Australia

© Cambridge University Press 1989

First published 1989
Reprinted 1989 (thrice), 1990

Printed in the United States of America

Library of Congress Cataloging-in-Publication Data
Rorty, Richard.
Contingency, irony, and solidarity / Richard Rorty.
p.   cm.
Includes index.
ISBN 0-521-35381-5. ISBN 0-521-36781-6 (pbk.)
1. Languages – Philosophy. I. Title.
P106.R586   1989   88–23358
401 – dc19   CIP

British Library Cataloging in Publication applied for

ISBN 0 521 35381 5 hardback
ISBN 0 521 36781 6 paperback

In memory of six liberals:
my parents and grandparents

The *agélastes* [Rabelais's word for those who do not laugh], the non-thought of received ideas, and kitsch are one and the same, the three-headed enemy of the art born as the echo of God's laughter, the art that created the fascinating imaginative realm where no one owns the truth and everyone has the right to be understood. That imaginative realm of tolerance was born with modern Europe, it is the very image of Europe – or at least our dream of Europe, a dream many times betrayed but nonetheless strong enough to unite us all in the fraternity that stretches far beyond the little European continent. But we know that the world where the individual is respected (the imaginative world of the novel, and the real one of Europe) is fragile and perishable. . . . if European culture seems under threat today, if the threat from within and without hangs over what is most precious about it – its respect for the individual, for his original thought, and for his right to an inviolable private life – then, I believe, that precious essence of the European spirit is being held safe as in a treasure chest inside the history of the novel, the wisdom of the novel.

Milan Kundera, *The Art of the Novel*

Contents

Preface

This book is based on two sets of lectures: three Northcliffe Lectures given at University College, London, in February of 1986 and four Clark Lectures given at Trinity College, Cambridge, in February of 1987. Slightly revised versions of the Northcliffe Lectures were published in the *London Review of Books* in the spring of 1986. They have been further revised to form the first three chapters of this book. A shortened version of Chapter 7, on Nabokov, was delivered as a Belitt Lecture at Bennington College and published by that college as a Bennington Chapbook on Literature. The other chapters have not been published previously.

Parts of this book skate on pretty thin ice – the passages in which I offer controversial interpretations of authors whom I discuss only briefly. This is particularly true of my treatment of Proust and of Hegel – authors about whom I hope someday to write more fully. But in other parts of the book the ice is a bit thicker. The footnotes in those parts cite my previous writings on various figures (e.g., Davidson, Dennett, Rawls, Freud, Heidegger, Derrida, Foucault, Habermas), writings which, I hope, back up some of the controversial things I say about them in this book. Most of the items cited will be reprinted in two volumes of my collected papers (provisionally entitled *Objectivity, Truth, and Relativism* and *Essays on Heidegger and Others*) to be published by Cambridge University Press.

I am very grateful to Karl Miller, Lord Northcliffe Professor of English Literature in University College and Editor of the *London Review of Books,* for his invitation to lecture at University College, as well as for encouragement and advice. I am equally grateful to the Master and Fellows of Trinity College, both for their invitation to give the Clark Lectures and for generous hospitality during my visit to Cambridge. I owe a great deal to the three institutions which granted me the leisure to compose these lectures: the John D. and Catherine T. MacArthur Foundation, the Center for Advanced Study of the University of Virginia, and the Wissenschaftskolleg zu Berlin. The MacArthur Fellowship which I held from 1981 to 1986 made it easy for me to branch out into new areas of reading and writing. The Director of the Center for Advanced Study, Dexter Whitehead, let me arrange my teaching in such a way as to maximize the opportunities provided by the MacArthur Fellowship. The patient and helpful staff of the Wissenschaftskolleg, surely the most supportive environment for scholars ever created, made my stay there in 1986–1987 both productive and pleasant.

As I revised and added to the two sets of lectures, gradually shaping this book, I received acute and helpful comments from friends who kindly took the time to read all or part of a growing pile of manuscript. Jeffrey Stout, David Bromwich, and Barry Allen saved me from many blunders and made a lot of useful suggestions. Konstantin Kolenda suggested a crucial rearrangement of topics. Charles Guignon, David Hiley, and Michael Levenson provided helpful bits of last-minute advice. I thank them all. I am grateful also to Eusebia Estes, Lyell Asher, and Meredith Garmon for secretarial and editorial assistance, and to Nancy Landau for careful copy-editing. Jeremy Mynott and Terence Moore, of Cambridge University Press, were constantly helpful and encouraging.

Introduction

The attempt to fuse the public and the private lies behind both Plato's attempt to answer the question "Why is it in one's interest to be just?" and Christianity's claim that perfect self-realization can be attained through service to others. Such metaphysical or theological attempts to unite a striving for perfection with a sense of community require us to acknowledge a common human nature. They ask us to believe that what is most important to each of us is what we have in common with others — that the springs of private fulfillment and of human solidarity are the same. Skeptics like Nietzsche have urged that metaphysics and theology are transparent attempts to make altruism look more reasonable than it is. Yet such skeptics typically have their own theories of human nature. They, too, claim that there is something common to all human beings — for example, the will to power, or libidinal impulses. Their point is that at the "deepest" level of the self there is *no* sense of human solidarity, that this sense is a "mere" artifact of human socialization. So such skeptics become antisocial. They turn their backs on the very idea of a community larger than a tiny circle of initiates.

Ever since Hegel, however, historicist thinkers have tried to get beyond this familiar standoff. They have denied that there is such a thing as "human nature" or the "deepest level of the self." Their strategy has been to insist that socialization, and thus historical circumstance, goes all the way down — that there is nothing "beneath" socialization or prior to history which is definatory of the human. Such writers tell us that the question "What is it to be a human being?" should be replaced by questions like "What is it to inhabit a rich twentieth-century democratic society?" and "How can an inhabitant of such a society be more than the enactor of a role in a previously written script?" This historicist turn has helped free us, gradually but steadily, from theology and metaphysics — from the temptation to look for an escape from time and chance. It has helped us substitute Freedom for Truth as the goal of thinking and of social progress. But even after this substitution takes place, the old tension between the private and the public remains. Historicists in whom the desire for self-creation, for private autonomy, dominates (e.g., Heidegger and Foucault) still tend to see socialization as Nietzsche did —

as antithetical to something deep within us. Historicists in whom the desire for a more just and free human community dominates (e.g., Dewey and Habermas) are still inclined to see the desire for private perfection as infected with "irrationalism" and "aestheticism." This book tries to do justice to both groups of historicist writers. I urge that we not try to choose between them but, rather, give them equal weight and then use them for different purposes. Authors like Kierkegaard, Nietzsche, Baudelaire, Proust, Heidegger, and Nabokov are useful as exemplars, as illustrations of what private perfection – a self-created, autonomous, human life – can be like. Authors such as Marx, Mill, Dewey, Habermas, and Rawls are fellow citizens rather than exemplars. They are engaged in a shared, social effort – the effort to make our institutions and practices more just and less cruel. We shall only think of these two kinds of writers as *opposed* if we think that a more comprehensive philosophical outlook would let us hold self-creation and justice, private perfection and human solidarity, in a single vision.

There is no way in which philosophy, or any other theoretical discipline, will ever let us do that. The closest we will come to joining these two quests is to see the aim of a just and free society as letting its citizens be as privatistic, "irrationalist," and aestheticist as they please so long as they do it on their own time – causing no harm to others and using no resources needed by those less advantaged. There are practical measures to be taken to accomplish this practical goal. But there is no way to bring self-creation together with justice at the level of theory. The vocabulary of self-creation is necessarily private, unshared, unsuited to argument. The vocabulary of justice is necessarily public and shared, a medium for argumentative exchange.

If we could bring ourselves to accept the fact that no theory about the nature of Man or Society or Rationality, or anything else, is going to synthesize Nietzsche with Marx or Heidegger with Habermas, we could begin to think of the relation between writers on autonomy and writers on justice as being like the relation between two kinds of tools – as little in need of synthesis as are paintbrushes and crowbars. One sort of writer lets us realize that the social virtues are not the only virtues, that some people have actually succeeded in re-creating themselves. We thereby become aware of our own half-articulate need to become a new person, one whom we as yet lack words to describe. The other sort reminds us of the failure of our institutions and practices to live up to the convictions to which we are already committed by the public, shared vocabulary we use in daily life. The one tells us that we need not speak only the language of the tribe, that we may find our own words, that we may have a responsibility to ourselves to find them. The other tells us that that responsibility is not the only one we have. Both are right, but there is no way to make both speak a single language.

This book tries to show how things look if we drop the demand for a theory which unifies the public and private, and are content to treat the demands of self-creation and of human solidarity as equally valid, yet forever incommensurable. It sketches a figure whom I call the "liberal ironist." I borrow my definition of "liberal" from Judith Shklar, who says that liberals are the people who think that cruelty is the worst thing we do. I use "ironist" to name the sort of person who faces up to the contingency of his or her own most central beliefs and desires – someone sufficiently historicist and nominalist to have abandoned the idea that those central beliefs and desires refer back to something beyond the reach of time and chance. Liberal ironists are people who include among these ungroundable desires their own hope that suffering will be diminished, that the humiliation of human beings by other human beings may cease.

For liberal ironists, there is no answer to the question "Why not be cruel?" – no noncircular theoretical backup for the belief that cruelty is horrible. Nor is there an answer to the question "How do you decide when to struggle against injustice and when to devote yourself to private projects of self-creation?" This question strikes liberal ironists as just as hopeless as the questions "Is it right to deliver n innocents over to be tortured to save the lives of $m \times n$ other innocents? If so, what are the correct values of n and m?" or the question "When may one favor members of one's family, or one's community, over other, randomly chosen, human beings?" Anybody who thinks that there are well-grounded theoretical answers to this sort of question – algorithms for resolving moral dilemmas of this sort – is still, in his heart, a theologian or a metaphysician. He believes in an order beyond time and change which both determines the point of human existence and establishes a hierarchy of responsibilities.

The ironist intellectuals who do not believe that there is such an order are far outnumbered (even in the lucky, rich, literate democracies) by people who believe that there *must* be one. Most nonintellectuals are still committed either to some form of religious faith or to some form of Enlightenment rationalism. So ironism has often seemed intrinsically hostile not only to democracy but to human solidarity – to solidarity with the mass of mankind, all those people who are convinced that such an order must exist. But it is not. Hostility to a particular historically conditioned and possibly transient form of solidarity is not hostility to solidarity as such. One of my aims in this book is to suggest the possibility of a liberal utopia: one in which ironism, in the relevant sense, is universal.

A postmetaphysical culture seems to me no more impossible than a postreligious one, and equally desirable.

In my utopia, human solidarity would be seen not as a fact to be recognized by clearing away "prejudice" or burrowing down to previously hidden depths but, rather, as a goal to be achieved. It is to be achieved not by inquiry but by imagination, the imaginative ability to see strange people as fellow sufferers. Solidarity is not discovered by reflection but created. It is created by increasing our sensitivity to the particular details of the pain and humiliation of other, unfamiliar sorts of people. Such increased sensitivity makes it more difficult to marginalize people different from ourselves by thinking, "They do not feel it as *we* would," or "There must always be suffering, so why not let *them* suffer?"

This process of coming to see other human beings as "one of us" rather than as "them" is a matter of detailed description of what unfamiliar people are like and of redescription of what we ourselves are like. This is a task not for theory but for genres such as ethnography, the journalist's report, the comic book, the docudrama, and, especially, the novel. Fiction like that of Dickens, Olive Schreiner, or Richard Wright gives us the details about kinds of suffering being endured by people to whom we had previously not attended. Fiction like that of Choderlos de Laclos, Henry James, or Nabokov gives us the details about what sorts of cruelty we ourselves are capable of, and thereby lets us redescribe ourselves. That is why the novel, the movie, and the TV program have, gradually but steadily, replaced the sermon and the treatise as the principal vehicles of moral change and progress.

In my liberal utopia, this replacement would receive a kind of recognition which it still lacks. That recognition would be part of a general turn against theory and toward narrative. Such a turn would be emblematic of our having given up the attempt to hold all the sides of our life in a single vision, to describe them with a single vocabulary. It would amount to a recognition of what, in Chapter 1, I call the "contingency of language" – the fact that there is no way to step outside the various vocabularies we have employed and find a metavocabulary which somehow takes account of *all possible* vocabularies, all possible ways of judging and feeling. A historicist and nominalist culture of the sort I envisage would settle instead for narratives which connect the present with the past, on the one hand, and with utopian futures, on the other. More important, it would regard the realization of utopias, and the envisaging of still further utopias, as an endless process – an endless, proliferating realization of Freedom, rather than a convergence toward an already existing Truth.

PART I

Contingency

I

The contingency of language

About two hundred years ago, the idea that truth was made rather than found began to take hold of the imagination of Europe. The French Revolution had shown that the whole vocabulary of social relations, and the whole spectrum of social institutions, could be replaced almost overnight. This precedent made utopian politics the rule rather than the exception among intellectuals. Utopian politics sets aside questions about both the will of God and the nature of man and dreams of creating a hitherto unknown form of society.

At about the same time, the Romantic poets were showing what happens when art is thought of no longer as imitation but, rather, as the artist's self-creation. The poets claimed for art the place in culture traditionally held by religion and philosophy, the place which the Enlightenment had claimed for science. The precedent the Romantics set lent initial plausibility to their claim. The actual role of novels, poems, plays, paintings, statues, and buildings in the social movements of the last century and a half has given it still greater plausibility.

By now these two tendencies have joined forces and have achieved cultural hegemony. For most contemporary intellectuals, questions of ends as opposed to means – questions about how to give a sense to one's own life or that of one's community – are questions for art or politics, or both, rather than for religion, philosophy, or science. This development has led to a split within philosophy. Some philosophers have remained faithful to the Enlightenment and have continued to identify themselves with the cause of science. They see the old struggle between science and religion, reason and unreason, as still going on, having now taken the form of a struggle between reason and all those forces within culture which think of truth as made rather than found. These philosophers take science as the paradigmatic human activity, and they insist that natural science discovers truth rather than makes it. They regard "making truth" as a merely metaphorical, and thoroughly misleading, phrase. They think of politics and art as spheres in which the notion of "truth" is out of place. Other philosophers, realizing that the world as it is described by the physical sciences teaches no moral lesson, offers no spiritual comfort, have concluded that science is no more than the handmaiden of technology. These philosophers have ranged themselves alongside the political utopian and the innovative artist.

Whereas the first kind of philosopher contrasts "hard scientific fact" with the "subjective" or with "metaphor," the second kind sees science as one more human activity, rather as the place at which human beings encounter a "hard," nonhuman reality. On this view, great scientists invent descriptions of the world which are useful for purposes of predicting and controlling what happens, just as poets and political thinkers invent other descriptions of it for other purposes. But there is no sense in which *any* of these descriptions is an accurate representation of the way the world is in itself. These philosophers regard the very idea of such a representation as pointless.

Had the first sort of philosopher, the sort whose hero is the natural scientist, always been the only sort, we should probably never have had an autonomous discipline called "philosophy" – a discipline as distinct from the sciences as it is from theology or from the arts. As such a discipline, philosophy is no more than two hundred years old. It owes its existence to attempts by the German idealists to put the sciences in their place and to give a clear sense to the vague idea that human beings make truth rather than find it. Kant wanted to consign science to the realm of second-rate truth – truth about a phenomenal world. Hegel wanted to think of natural science as a description of spirit not yet fully conscious of its own spiritual nature, and thereby to elevate the sort of truth offered by the poet and the political revolutionary to first-rate status.

German idealism, however, was a short-lived and unsatisfactory compromise. For Kant and Hegel went only halfway in their repudiation of the idea that truth is "out there." They were willing to view the world of empirical science as a made world – to see matter as constructed by mind, or as consisting in mind insufficiently conscious of its own mental character. But they persisted in seeing mind, spirit, the depths of the human self, as having an intrinsic nature – one which could be known by a kind of nonempirical super science called philosophy. This meant that only half of truth – the bottom, scientific half – was made. Higher truth, the truth about mind, the province of philosophy, was still a matter of discovery rather than creation.

What was needed, and what the idealists were unable to envisage, was a repudiation of the very idea of anything – mind or matter, self or world – having an intrinsic nature to be expressed or represented. For the idealists confused the idea that nothing has such a nature with the idea that space and time are unreal, that human beings cause the spatiotemporal world to exist.

We need to make a distinction between the claim that the world is out

4

there and the claim that truth is out there. To say that the world is out there, that it is not our creation, is to say, with common sense, that most things in space and time are the effects of causes which do not include human mental states. To say that truth is not out there is simply to say that where there are no sentences there is no truth, that sentences are elements of human languages, and that human languages are human creations.

Truth cannot be out there – cannot exist independently of the human mind – because sentences cannot so exist, or be out there. The world is out there, but descriptions of the world are not. Only descriptions of the world can be true or false. The world on its own – unaided by the describing activities of human beings – cannot.

The suggestion that truth, as well as the world, is out there is a legacy of an age in which the world was seen as the creation of a being who had a language of his own. If we cease to attempt to make sense of the idea of such a nonhuman language, we shall not be tempted to confuse the platitude that the world may cause us to be justified in believing a sentence true with the claim that the world splits itself up, on its own initiative, into sentence-shaped chunks called "facts." But if one clings to the notion of self-subsistent facts, it is easy to start capitalizing the word "truth" and treating it as something identical either with God or with the world as God's project. Then one will say, for example, that Truth is great, and will prevail.

This conflation is facilitated by confining attention to single sentences as opposed to vocabularies. For we often let the world decide the competition between alternative sentences (e.g., between "Red wins" and "Black wins" or between "The butler did it" and "The doctor did it"). In such cases, it is easy to run together the fact that the world contains the causes of our being justified in holding a belief with the claim that some nonlinguistic state of the world is itself an example of truth, or that some such state "makes a belief true" by "corresponding" to it. But it is not so easy when we turn from individual sentences to vocabularies as wholes. When we consider examples of alternative language games – the vocabulary of ancient Athenian politics versus Jefferson's, the moral vocabulary of Saint Paul versus Freud's, the jargon of Newton versus that of Aristotle, the idiom of Blake versus that of Dryden – it is difficult to think of the world as making one of these better than another, of the world as deciding between them. When the notion of "description of the world" is moved from the level of criterion-governed sentences within language games to language games as wholes, games which we do not choose between by reference to criteria, the idea that the world decides which descriptions are true can no longer be given a clear sense. It becomes hard to think that that vocabulary is somehow already out there in the world, waiting for us to discover it. Attention (of the sort fostered by intellectual historians like Thomas Kuhn and Quentin Skinner) to the vocabularies in which sentences are formulated, rather than to individual sentences, makes us realize, for example, that the fact that Newton's vocabulary lets us predict the world more easily than Aristotle's does not mean that the world speaks Newtonian.

The world does not speak. Only we do. The world can, once we have programmed ourselves with a language, cause us to hold beliefs. But it cannot propose a language for us to speak. Only other human beings can do that. The realization that the world does not tell us what language games to play should not, however, lead us to say that a decision about which to play is arbitrary, nor to say that it is the expression of something deep within us. The moral is not that objective criteria for choice of vocabulary are to be replaced with subjective criteria, reason with will or feeling. It is rather that the notions of criteria and choice (including that of "arbitrary" choice) are no longer in point when it comes to changes from one language game to another. Europe did not *decide* to accept the idiom of Romantic poetry, or of socialist politics, or of Galilean mechanics. That sort of shift was no more an act of will than it was a result of argument. Rather, Europe gradually lost the habit of using certain words and gradually acquired the habit of using others.

As Kuhn argues in *The Copernican Revolution,* we did not decide on the basis of some telescopic observations, or on the basis of anything else, that the earth was not the center of the universe, that macroscopic behavior could be explained on the basis of microstructural motion, and that prediction and control should be the principal aim of scientific theorizing. Rather, after a hundred years of inconclusive muddle, the Europeans found themselves speaking in a way which took these interlocked theses for granted. Cultural change of this magnitude does not result from applying criteria (or from "arbitrary decision") any more than individuals become theists or atheists, or shift from one spouse or circle of friends to another, as a result either of applying criteria or of *actes gratuits.* We should not look within ourselves for criteria of decision in such matters any more than we should look to the world.

The temptation to look for criteria is a species of the more general temptation to think of the world, or the human self, as possessing an intrinsic nature, an essence. That is, it is the result of the temptation to privilege some one among the many languages in which we habitually describe the world or ourselves. As long as we think that there is some relation called "fitting the world" or "expressing the real nature of the self" which can be possessed or lacked by vocabularies-as-wholes, we

6

shall continue the traditional philosophical search for a criterion to tell us which vocabularies have this desirable feature. But if we could ever become reconciled to the idea that most of reality is indifferent to our descriptions of it, and that the human self is created by the use of a vocabulary rather than being adequately or inadequately expressed in a vocabulary, then we should at last have assimilated what was true in the Romantic idea that truth is made rather than found. What is true about this claim is just that *languages* are made rather than found, and that truth is a property of linguistic entities, of sentences.[1]

I can sum up by redescribing what, in my view, the revolutionaries and poets of two centuries ago were getting at. What was glimpsed at the end of the eighteenth century was that anything could be made to look good or bad, important or unimportant, useful or useless, by being redescribed. What Hegel describes as the process of spirit gradually becoming self-conscious of its intrinsic nature is better described as the process of European linguistic practices changing at a faster and faster rate. The phenomenon Hegel describes is that of more people offering more radical redescriptions of more things than ever before, of young people going through half a dozen spiritual gestalt-switches before reaching adulthood. What the Romantics expressed as the claim that imagination, rather than reason, is the central human faculty was the realization that a talent for speaking differently, rather than for arguing well, is the chief instrument of cultural change. What political utopians since the French Revolution have sensed is not that an enduring, substratal human nature has been suppressed or repressed by "unnatural" or "irrational" social institutions but rather that changing languages and other social practices may produce human beings of a sort that had never before existed. The German idealists, the French revolutionaries, and the Romantic poets had in common a dim sense that human beings whose language changed so that they no longer spoke of themselves as responsible to nonhuman powers would thereby become a new kind of human beings.

The difficulty faced by a philosopher who, like myself, is sympathetic

---

1 I have no criterion of individuation for distinct languages or vocabularies to offer, but I am not sure that we need one. Philosophers have used phrases like "in the language L" for a long time without worrying too much about how one can tell where one natural language ends and another begins, nor about when "the scientific vocabulary of the sixteenth century" ends and "the vocabulary of the New Science" begins. Roughly, a break of this sort occurs when we start using "translation" rather than "explanation" in talking about geographical or chronological differences. This will happen whenever we find it handy to start mentioning words rather than using them − to highlight the difference between two sets of human practices by putting quotation marks around elements of those practices.

to this suggestion – one who thinks of himself as auxiliary to the poet rather than to the physicist – is to avoid hinting that this suggestion gets something right, that my sort of philosophy corresponds to the way things really are. For this talk of correspondence brings back just the idea my sort of philosopher wants to get rid of, the idea that the world or the self has an intrinsic nature. From our point of view, explaining the success of science, or the desirability of political liberalism, by talk of "fitting the world" or "expressing human nature" is like explaining why opium makes you sleepy by talking about its dormitive power. To say that Freud's vocabulary gets at the truth about human nature, or Newton's at the truth about the heavens, is not an explanation of anything. It is just an empty compliment – one traditionally paid to writers whose novel jargon we have found useful. To say that there is no such thing as intrinsic nature is not to say that the intrinsic nature of reality has turned out, surprisingly enough, to be extrinsic. It is to say that the term "intrinsic nature" is one which it would pay us not to use, an expression which has caused more trouble than it has been worth. To say that we should drop the idea of truth as out there waiting to be discovered is not to say that we have discovered that, out there, there is no truth.[2] It is to say that our purposes would be served best by ceasing to see truth as a deep matter, as a topic of philosophical interest, or "true" as a term which repays "analysis." "The nature of truth" is an unprofitable topic, resembling in this respect "the nature of man" and "the nature of God," and differing from "the nature of the positron," and "the nature of Oedipal fixation." But this claim about relative profitability, in turn, is just the recommendation that we in fact *say* little about these topics, and see how we get on.

On the view of philosophy which I am offering, philosophers should not be asked for arguments against, for example, the correspondence theory of truth or the idea of the "intrinsic nature of reality." The trouble with arguments against the use of a familiar and time-honored vocabulary is that they are expected to be phrased in that very vocabulary. They are expected to show that central elements in that vocabulary are "inconsistent in their own terms" or that they "deconstruct themselves." But that can *never* be shown. Any argument to the effect that our familiar use of a familiar term is incoherent, or empty, or confused, or vague, or "merely

---

2 Nietzsche has caused a lot of confusion by inferring from "truth is not a matter of correspondence to reality" to "what we call 'truths' are just useful lies." The same confusion is occasionally found in Derrida, in the inference from "there is no such reality as the metaphysicians have hoped to find" to "what we call 'real' is not really real." Such confusions make Nietzsche and Derrida liable to charges of self-referential inconsistency – to claiming to know what they themselves claim cannot be known.

metaphorical" is bound to be inconclusive and question-begging. For such use is, after all, the paridigm of coherent, meaningful, literal, speech. Such arguments are always parasitic upon, and abbreviations for, claims that a better vocabulary is available. Interesting philosophy is rarely an examination of the pros and cons of a thesis. Usually it is, implicitly or explicitly, a contest between an entrenched vocabulary which has become a nuisance and a half-formed new vocabulary which vaguely promises great things.

The latter "method" of philosophy is the same as the "method" of utopian politics or revolutionary science (as opposed to parliamentary politics, or normal science). The method is to redescribe lots and lots of things in new ways, until you have created a pattern of linguistic behavior which will tempt the rising generation to adopt it, thereby causing them to look for appropriate new forms of nonlinguistic behavior, for example, the adoption of new scientific equipment or new social institutions. This sort of philosophy does not work piece by piece, analyzing concept after concept, or testing thesis after thesis. Rather, it works holistically and pragmatically. It says things like "try thinking of it this way" – or more specifically, "try to ignore the apparently futile traditional questions by substituting the following new and possibly interesting questions." It does not pretend to have a better candidate for doing the same old things which we did when we spoke in the old way. Rather, it suggests that we might want to stop doing those things and do something else. But it does not argue for this suggestion on the basis of antecedent criteria common to the old and the new language games. For just insofar as the new language really is new, there will be no such criteria.

Conforming to my own precepts, I am not going to offer arguments against the vocabulary I want to replace. Instead, I am going to try to make the vocabulary I favor look attractive by showing how it may be used to describe a variety of topics. More specifically, in this chapter I shall be describing the work of Donald Davidson in philosophy of language as a manifestation of a willingness to drop the idea of "intrinsic nature," a willingness to face up to the *contingency* of the language we use. In subsequent chapters, I shall try to show how a recognition of that contingency leads to a recognition of the contingency of conscience, and how both recognitions lead to a picture of intellectual and moral progress as a history of increasingly useful metaphors rather than of increasing understanding of how things really are.

I begin, in this first chapter, with the philosophy of language because I want to spell out the consequences of my claims that only sentences can be true, and that human beings make truths by making languages in which to phrase sentences. I shall concentrate on the work of Davidson because he is the philosopher who has done most to explore these consequences.[3] Davidson's treatment of truth ties in with his treatment of language learning and of metaphor to form the first systematic treatment of language which breaks *completely* with the notion of language as something which can be adequate or inadequate to the world or to the self. For Davidson breaks with the notion that language is a *medium* – a medium either of representation or of expression.

I can explain what I mean by a medium by noting that the traditional picture of the human situation has been one in which human beings are not simply networks of beliefs and desires but rather beings which *have* those beliefs and desires. The traditional view is that there is a core self which can look at, decide among, use, and express itself by means of, such beliefs and desires. Further, these beliefs and desires are criticizable not simply by reference to their ability to cohere with one another, but by reference to something exterior to the network within which they are strands. Beliefs are, on this account, criticizable because they fail to correspond to reality. Desires are criticizable because they fail to correspond to the essential nature of the human self – because they are "irrational" or "unnatural." So we have a picture of the essential core of the self on one side of this network of beliefs and desires, and reality on the other side. In this picture, the network is the product of an interaction between the two, alternately expressing the one and representing the other. This is the traditional subject-object picture which idealism tried and failed to replace, and which Nietzsche, Heidegger, Derrida, James, Dewey, Goodman, Sellars, Putnam, Davidson and others have tried to replace without entangling themselves in the idealists' paradoxes.

One phase of this effort of replacement consisted in an attempt to substitute "language" for "mind" or "consciousness" as the medium out of which beliefs and desires are constructed, the third, mediating, element between self and world. This turn toward language was thought of as a progressive, naturalizing move. It seemed so because it seemed easier to give a causal account of the evolutionary emergence of language-using organisms than of the metaphysical emergence of consciousness out of nonconsciousness. But in itself this substitution is ineffective. For if we stick to the picture of language as a medium, something

---

3 I should remark that Davidson cannot be held responsible for the interpretation I am putting on his views, nor for the further views I extrapolate from his. For an extended statement of that interpretation, see my "Pragmatism, Davidson and Truth," in Ernest Lepore, ed., *Truth and Interpretation: Perspectives on the Philosophy of Donald Davidson* (Oxford: Blackwell, 1984). For Davidson's reaction to this interpretation, see his "After-thoughts" to "A Coherence Theory of Truth and Knowledge," in Alan Malachowski, *Reading Rorty* (Oxford: Blackwell, in press).

standing between the self and the nonhuman reality with which the self seeks to be in touch, we have made no progress. We are still using a subject-object picture, and we are still stuck with issues about skepticism, idealism, and realism. For we are still able to ask questions about language of the same sort we asked about consciousness.

These are such questions as: "Does the medium between the self and reality get them together or keep them apart?" "Should we see the medium primarily as a medium of expression – of articulating what lies deep within the self? Or should we see it as primarily a medium of representation – showing the self what lies outside it?" Idealist theories of knowledge and Romantic notions of the imagination can, alas, easily be transposed from the jargon of "consciousness" into that of "language." Realistic and moralistic reactions to such theories can be transposed equally easily. So the seesaw battles between romanticism and moralism, and between idealism and realism, will continue as long as one thinks there is a hope of making sense of the question of whether a given language is "adequate" to a task – either the task of properly expressing the nature of the human species, or the task of properly representing the structure of nonhuman reality.

We need to get off this seesaw. Davidson helps us do so. For he does not view language as a medium for either expression or representation. So he is able to set aside the idea that both the self and reality have intrinsic natures, natures which are out there waiting to be known. Davidson's view of language is neither reductionist nor expansionist. It does not, as analytical philosophers sometimes have, purport to give reductive definitions of semantical notions like "truth" or "intentionality" or "reference." Nor does it resemble Heidegger's attempt to make language into a kind of divinity, something of which human beings are mere emanations. As Derrida has warned us, such an apotheosis of language is merely a transposed version of the idealists' apotheosis of consciousness.

In avoiding both reductionism and expansionism, Davidson resembles Wittgenstein. Both philosophers treat alternative vocabularies as more like alternative tools than like bits of a jigsaw puzzle. To treat them as pieces of a puzzle is to assume that all vocabularies are dispensable, or reducible to other vocabularies, or capable of being united with all other vocabularies in one grand unified super vocabulary. If we avoid this assumption, we shall not be inclined to ask questions like "What is the place of consciousness in a world of molecules?" "Are colors more mind-dependent than weights?" "What is the place of value in a world of fact?" "What is the place of intentionality in a world of causation?" "What is the relation between the solid table of common sense and the unsolid table of microphysics?" or "What is the relation of language to thought?" We should not try to answer such questions, for doing so leads either to the evident failures of reductionism or to the short-lived successes of expansionism. We should restrict ourselves to questions like "Does our use of these words get in the way of our use of those other words?" This is a question about whether our use of tools is inefficient, not a question about whether our beliefs are contradictory.

"Merely philosophical" questions, like Eddington's question about the two tables, are attempts to stir up a factitious theoretical quarrel between vocabularies which have proved capable of peaceful coexistence. The questions I have recited above are all cases in which philosophers have given their subject a bad name by seeing difficulties nobody else sees. But this is not to say that vocabularies never do get in the way of each other. On the contrary, revolutionary achievements in the arts, in the sciences, and in moral and political thought typically occur when somebody realizes that two or more of our vocabularies are interfering with each other, and proceeds to invent a new vocabulary to replace both. For example, the traditional Aristotelian vocabulary got in the way of the mathematized vocabulary that was being developed in the sixteenth century by students of mechanics. Again, young German theology students of the late eighteenth century – like Hegel and Hölderlin – found that the vocabulary in which they worshiped Jesus was getting in the way of the vocabulary in which they worshiped the Greeks. Yet again, the use of Rossetti-like tropes got in the way of the early Yeats's use of Blakean tropes.

The gradual trial-and-error creation of a new, third, vocabulary – the sort of vocabulary developed by people like Galileo, Hegel, or the later Yeats – is not a discovery about how old vocabularies fit together. That is why it cannot be reached by an inferential process – by starting with premises formulated in the old vocabularies. Such creations are not the result of successfully fitting together pieces of a puzzle. They are not discoveries of a reality behind the appearances, of an undistorted view of the whole picture with which to replace myopic views of its parts. The proper analogy is with the invention of new tools to take the place of old tools. To come up with such a vocabulary is more like discarding the lever and the chock because one has envisaged the pully, or like discarding gesso and tempera because one has now figured out how to size canvas properly.

This Wittgensteinian analogy between vocabularies and tools has one obvious drawback. The craftsman typically knows what job he needs to do before picking or inventing tools with which to do it. By contrast, someone like Galileo, Yeats, or Hegel (a "poet" in my wide sense of the term – the sense of "one who makes things new") is typically unable to make clear exactly what it is that he wants to do before developing the language in which he succeeds in doing it. His new vocabulary makes possible, for the first time, a formulation of its own purpose. It is a tool for doing something which could not have been envisaged prior to the development of a particular set of descriptions, those which it itself helps to provide. But I shall, for the moment, ignore this disanalogy. I want simply to remark that the contrast between the jigsaw-puzzle and the "tool" models of alternative vocabularies reflects the contrast between – in Nietzsche's slightly misleading terms – the will to truth and the will to self-overcoming. Both are expressions of the contrast between the attempt to represent or express something that was already there and the attempt to make something that never had been dreamed of before.

Davidson spells out the implications of Wittgenstein's treatment of vocabularies as tools by raising explicit doubts about the assumptions underlying traditional pre-Wittgensteinian accounts of language. These accounts have taken for granted that questions like "Is the language we are presently using the 'right' language – is it adequate to its task as a medium of expression or representation?" "Is our language a transparent or an opaque medium?" make sense. Such questions assume there are relations such as "fitting the world" or "being faithful to the true nature of the self" in which language might stand to nonlanguage. This assumption goes along with the assumption that "our language" – the language we speak now, the vocabulary at the disposal of educated inhabitants of the twentieth century – is somehow a unity, a third thing which stands in some determinate relation with two other unities – the self and reality. Both assumptions are natural enough, once we accept the idea that there are nonlinguistic things called "meanings" which it is the task of language to express, as well as the idea that there are nonlinguistic things called "facts" which it is the task of language to represent. Both ideas enshrine the notion of language as medium.

Davidson's polemics against the traditional philosophical uses of the terms "fact" and "meaning," and against what he calls "the scheme-content model" of thought and inquiry, are parts of a larger polemic against the idea that there is a fixed task for language to perform, and an entity called "language" or "the language" or "our language" which may or may not be performing this task efficiently. Davidson's doubt that there is any such entity parallels Gilbert Ryle's and Daniel Dennett's doubts about whether there is anything called "the mind" or "consciousness."[4] Both sets of doubts are doubts about the utility of the notion of a

---

4 For an elaboration of these doubts, see my "Contemporary Philosophy of Mind,"

medium between the self and reality – the sort of medium which realists see as transparent and skeptics as opaque.

In a recent paper, nicely entitled "A Nice Derangement of Epitaphs,"[5] Davidson tries to undermine the notion of languages as entities by developing the notion of what he calls "a passing theory" about the noises and inscriptions presently being produced by a fellow human. Think of such a theory as part of a larger "passing theory" about this person's total behavior – a set of guesses about what she will do under what conditions. Such a theory is "passing" because it must constantly be corrected to allow for mumbles, stumbles, malapropisms, metaphors, tics, seizures, psychotic symptoms, egregious stupidity, strokes of genius, and the like. To make things easier, imagine that I am forming such a theory about the current behavior of a native of an exotic culture into which I have unexpectedly parachuted. This strange person, who presumably finds me equally strange, will simultaneously be busy forming a theory about my behavior. If we ever succeed in communicating easily and happily, it will be because her guesses about what I am going to do next, including what noises I am going to make next, and my own expectations about what I shall do or say under certain circumstances, come more or less to coincide, and because the converse is also true. She and I are coping with each other as we might cope with mangoes or boa constrictors – we are trying not to be taken by surprise. To say that we come to speak the same language is to say, as Davidson puts it, that "we tend to converge on passing theories." Davidson's point is that all "two people need, if they are to understand one another through speech, is the ability to converge on passing theories from utterance to utterance."

Davidson's account of linguistic communication dispenses with the picture of language as a third thing intervening between self and reality, and of different languages as barriers between persons or cultures. To say that one's previous language was inappropriate for dealing with some segment of the world (for example, the starry heavens above, or the raging passions within) is just to say that one is now, having learned a new language, able to handle that segment more easily. To say that two communities have trouble getting along because the words they use are so hard to translate into each other is just to say that the linguistic behavior of inhabitants of one community may, like the rest of their behavior, be hard for inhabitants of the other community to predict. As Davidson puts it,

Synthese 53 (1982): 332–348. For Dennett's doubts about my interpretations of his views, see his "Comments on Rorty," pp. 348–354.
5   This essay can be found in Lepore, ed., *Truth and Interpretation.*

We should realize that we have abandoned not only the ordinary notion of a language, but we have erased the boundary between knowing a language and knowing our way around the world generally. For there are no rules for arriving at passing theories that work. . . . There is no more chance of regularizing, or teaching, this process than there is of regularizing or teaching the process of creating new theories to cope with new data – for that is what this process involves. . . .

There is no such thing as a language, not if a language is anything like what philosophers, at least, have supposed. There is therefore no such thing to be learned or mastered. We must give up the idea of a clearly defined shared structure which language users master and then apply to cases . . . We should give up the attempt to illuminate how we communicate by appeal to conventions.[6]

This line of thought about language is analogous to the Ryle-Dennett view that when we use a mentalistic terminology we are simply using an efficient vocabulary – the vocabulary characteristic of what Dennett calls the "intentional stance" – to predict what an organism is likely to do or say under various sets of circumstances. Davidson is a nonreductive behaviorist about language in the same way that Ryle was a nonreductive behaviorist about mind. Neither has any desire to give equivalents in Behaviorese for talk about beliefs or about reference. But both are saying: Think of the term "mind" or "language" not as the name of a medium between self and reality but simply as a flag which signals the desirability of using a certain vocabulary when trying to cope with certain kinds of organisms. To say that a given organism – or, for that matter, a given machine – has a mind is just to say that, for some purposes, it will pay to think of it as having beliefs and desires. To say that it is a language user is just to say that pairing off the marks and noises it makes with those we make will prove a useful tactic in predicting and controlling its future behavior.

This Wittgensteinian attitude, developed by Ryle and Dennett for minds and by Davidson for languages, naturalizes mind and language by making all questions about the relation of either to the rest of universe *causal* questions, as opposed to questions about adequacy of representation or expression. It makes perfectly good sense to ask how we got from the relative mindlessness of the monkey to the full-fledged mindedness of the human, or from speaking Neanderthal to speaking postmodern, if these are construed as straightforward causal questions. In the former case the answer takes us off into neurology and thence into evolutionary

6   "A Nice Derangement of Epitaphs," in Lepore, ed., *Truth and Interpretation*, p. 446. Italics added.

biology. But in the latter case it takes us into intellectual history viewed as the history of metaphor. For my purposes in this book, it is the latter which is important. So I shall spend the rest of this chapter sketching an account of intellectual and moral progress which squares with Davidson's account of language.

To see the history of language, and thus of the arts, the sciences, and the moral sense, as the history of metaphor is to drop the picture of the human mind, or human languages, becoming better and better suited to the purposes for which God or Nature designed them, for example, able to express more and more meanings or to represent more and more facts. The idea that language has a purpose goes once the idea of language as medium goes. A culture which renounced both ideas would be the triumph of those tendencies in modern thought which began two hundred years ago, the tendencies common to German idealism, Romantic poetry, and utopian politics.

A nonteleological view of intellectual history, including the history of science, does for the theory of culture what the Mendelian, mechanistic, account of natural selection did for evolutionary theory. Mendel let us see mind as something which just happened rather than as something which was the point of the whole process. Davidson lets us think of the history of language, and thus of culture, as Darwin taught us to think of the history of a coral reef. Old metaphors are constantly dying off into literalness, and then serving as a platform and foil for new metaphors. This analogy lets us think of "our language" – that is, of the science and culture of twentieth-century Europe – as something that took shape as a result of a great number of sheer contingencies. Our language and our culture are as much a contingency, as much a result of thousands of small mutations finding niches (and millions of others finding no niches), as are the orchids and the anthropoids.

To accept this analogy, we must follow Mary Hesse in thinking of scientific revolutions as "metaphoric redescriptions" of nature rather than insights into the intrinsic nature of nature.[7] Further, we must resist the temptation to think that the redescriptions of reality offered by contemporary physical or biological science are somehow closer to "the things themselves," less "mind-dependent," than the redescriptions of history offered by contemporary culture criticism. We need to see the constellations of causal forces which produced talk of DNA or of the Big Bang as of a piece with the causal forces which produced talk of "secu-

7   See "The Explanatory Function of Metaphor," in Hesse, *Revolutions and Reconstructions in the Philosophy of Science* (Bloomington: Indiana University Press, 1980).

larization" or of "late capitalism."[8] These various constellations are the random factors which have made some things subjects of conversation for us and others not, have made some projects and not others possible and important.

I can develop the contrast between the idea that the history of culture has a *telos* – such as the discovery of truth, or the emancipation of humanity – and the Nietzschean and Davidsonian picture which I am sketching by noting that the latter picture is compatible with a bleakly mechanical description of the relation between human beings and the rest of the universe. For genuine novelty can, after all, occur in a world of blind, contingent, mechanical forces. Think of novelty as the sort of thing which happens when, for example, a cosmic ray scrambles the atoms in a DNA molecule, thus sending things off in the direction of the orchids or the anthropoids. The orchids, when their time came, were no less novel or marvelous for the sheer contingency of this necessary condition of their existence. Analogously, for all we know, or should care, Aristotle's metaphorical use of *ousia*, Saint Paul's metaphorical use of *agapē*, and Newton's metaphorical use of *gravitas*, were the results of cosmic rays scrambling the fine structure of some crucial neurons in their respective brains. Or, more plausibly, they were the result of some odd episodes in infancy – some obsessional kinks left in these brains by idiosyncratic traumata. It hardly matters how the trick was done. The results were marvelous. There had never been such things before.

This account of intellectual history chimes with Nietzsche's definition of "truth" as "a mobile army of metaphors." It also chimes with the description I offered earlier of people like Galileo and Hegel and Yeats, people in whose minds new vocabularies developed, thereby equipping them with tools for doing things which could not even have been envisaged before these tools were available. But in order to accept this picture, we need to see the distinction between the literal and the metaphorical in the way Davidson sees it: not as a distinction between two sorts of meaning, nor as a distinction between two sorts of interpretation, but as a distinction between familiar and unfamiliar uses of noises and marks. The literal uses of noises and marks are the uses we can handle by our old theories about what people will say under various conditions. Their metaphorical use is the sort which makes us get busy developing a new theory.

8 This coalescence is resisted in Bernard Williams's discussion of Davidson's and my views in chap. 6 of his *Ethics and the Limits of Philosophy* (Cambridge, Mass.: Harvard University Press, 1985). For a partial reply to Williams, see my "Is Natural Science a Natural Kind?" in Ernan McMullin, ed., *Construction and Constraint: The Shaping of Scientific Rationality* (Notre Dame, Ind.: University of Notre Dame Press, 1988).

Davidson puts this point by saying that one should not think of meta-phorical expressions as having meanings distinct from their literal ones. To have a meaning is to have a place in a language game. Metaphors, by definition, do not. Davidson denies, in his words, "the thesis that associ-ated with a metaphor is a cognitive content that its author wishes to convey and that the interpreter must grasp if he is to get the message."[9] In his view, tossing a metaphor into a conversation is like suddenly breaking off the conversation long enough to make a face, or pulling a photograph out of your pocket and displaying it, or pointing at a feature of the surroundings, or slapping your interlocutor's face, or kissing him. Tossing a metaphor into a text is like using italics, or illustrations, or odd punctuation or formats.

All these are ways of producing effects on your interlocutor or your reader, but not ways of conveying a message. To none of these is it appropriate to respond with "What exactly are you trying to say?" If one had wanted to say something – if one had wanted to utter a sentence with a meaning – one would presumably have done so. But instead one thought that one's aim could be better carried out by other means. That one uses familiar words in unfamiliar ways – rather than slaps, kisses, pictures, gestures, or grimaces – does not show that what one said must have a meaning. An attempt to state that meaning would be an attempt to find some familiar (that is, literal) use of words – some sentence which already had a place in the language game – and, to claim that one might just as well have *that*. But the unparaphrasability of metaphor is just the unsuitability of any such familiar sentence for one's purpose.

Uttering a sentence without a fixed place in a language game is, as the positivists rightly have said, to utter something which is neither true nor false – something which is not, in Ian Hacking's terms, a "truth-value candidate." This is because it is a sentence which one cannot confirm or disconfirm, argue for or against. One can only savor it or spit it out. But this is not to say that it may not, in time, *become* a truth-value candidate. If it *is* savored rather than spat out, the sentence may be repeated, caught up, bandied about. Then it will gradually require a habitual use, a familiar place in the language game. It will thereby have ceased to be a metaphor – or, if you like, it will have become what most sentences of our language are, a dead metaphor. It will be just one more, literally true or literally false, sentence of the language. That is to say, our theories about the linguistic behavior of our fellows will suffice to let us cope with its

9 Davidson, "What Metaphors Mean," in his *Inquiries into Truth and Interpretation* (Oxford University Press, 1984), p. 262.

utterance in the same unthinking way in which we cope with most of their other utterances.

The Davidsonian claim that metaphors do not have meanings may seem like a typical philosopher's quibble, but it is not.[10] It is part of an attempt to get us to stop thinking of language as a medium. This, in turn, is part of a larger attempt to get rid of the traditional philosophical picture of what it is to be human. The importance of Davidson's point can perhaps best be seen by contrasting his treatment of metaphor with those of the Platonist and the positivist on the one hand and the Romantic on the other. The Platonist and the positivist share a reductionist view of metaphor: They think metaphors are either paraphrasable or useless for the one serious purpose which language has, namely, representing reality. By contrast, the Romantic has an expansionist view: He thinks metaphor is strange, mystic, wonderful. Romantics attribute metaphor to a mysterious faculty called the "imagination," a faculty they suppose to be at the very center of the self, the deep heart's core. Whereas the metaphorical looks irrelevant to Platonists and positivists, the literal looks irrelevant to Romantics. For the former think that the point of language is to represent a hidden reality which lies outside us, and the latter thinks its purpose is to express a hidden reality which lies within us.

Positivist history of culture thus sees language as gradually shaping itself around the contours of the physical world. Romantic history of culture sees language as gradually bringing Spirit to self-consciousness. Nietzschean history of culture, and Davidsonian philosophy of language, see language as we now see evolution, as new forms of life constantly killing off old forms – not to accomplish a higher purpose, but blindly. Whereas the positivist sees Galileo as making a discovery – finally coming up with the words which were needed to fit the world properly, words Aristotle missed – the Davidsonian sees him as having hit upon a tool which happened to work better for certain purposes than any previous tool. Once we found out what could be done with a Galilean vocabulary, nobody was much interested in doing the things which used to be done (and which Thomists thought should still be done) with an Aristotelian vocabulary.

Similarly, whereas the Romantic sees Yeats as having gotten at something which nobody had previously gotten at, expressed something which had long been yearning for expression, the Davidsonian sees him as having hit upon some tools which enabled him to write poems which

---

10 For a further defense of Davidson against the charge of quibbling, and various other charges, see my "Unfamiliar Noises: Hesse and Davidson on Metaphor," *Proceedings of the Aristotelian Society*, supplementary vol. 61 (1987): 283–296.

were not just variations on the poems of his precursors. Once we had Yeats's later poems in hand, we were less interested in reading Rossetti's. What goes for revolutionary, strong scientists and poets goes also for strong philosophers – people like Hegel and Davidson, the sort of philosophers who are interested in dissolving inherited problems rather than in solving them. In this view, substituting dialectic for demonstration as the method of philosophy, or getting rid of the correspondence theory of truth, is not a discovery about the nature of a preexistent entity called "philosophy" or "truth." It is changing the way we talk, and thereby changing what we want to do and what we think we are.

But in a Nietzschean view, one which drops the reality-appearance distinction, to change how we talk is to change what, for our own purposes, we are. To say, with Nietzsche, that God is dead, is to say that we serve no higher purposes. The Nietzschean substitution of self-creation for discovery substitutes a picture of the hungry generations treading each other down for a picture of humanity approaching closer and closer to the light. A culture in which Nietzschean metaphors were literalized would be one which took for granted that philosophical problems are as temporary as poetic problems, that there are no problems which bind the generations together into a single natural kind called "humanity." A sense of human history as the history of successive metaphors would let us see the poet, in the generic sense of the maker of new words, the shaper of new languages, as the vanguard of the species.

I shall try to develop this last point in Chapters 2 and 3 in terms of Harold Bloom's notion of the "strong poet." But I shall end this first chapter by going back to the claim, which has been central to what I have been saying, that the world does not provide us with any criterion of choice between alternative metaphors, that we can only compare languages or metaphors with one another, not with something beyond language called "fact."

The only way to argue for this claim is to do what philosophers like Goodman, Putnam, and Davidson have done: exhibit the sterility of attempts to give a sense to phrases like "the way the world is" or "fitting the facts." Such efforts can be supplemented by the work of philosophers of science such as Kuhn and Hesse. These philosophers explain why there is no way to explain the fact that a Galilean vocabulary enables us to make better predictions than an Aristotelian vocabulary by the claim that the book of nature is written in the language of mathematics.

These sorts of arguments by philosophers of language and of science should be seen against the background of the work of intellectual historians: historians who, like Hans Blumenberg, have tried to trace the similarities and dissimilarities between the Age of Faith and the Age of Reason.[11] These historians have made the point I mentioned earlier: The very idea that the world or the self has an intrinsic nature – one which the physicist or the poet may have glimpsed – is a remnant of the idea that the world is a divine creation, the work of someone who had something in mind, who Himself spoke some language in which He described His own project. Only if we have some such picture in mind, some picture of the universe as either itself a person or as created by a person, can we make sense of the idea that the world has an "intrinsic nature." For the cash value of that phrase is just that some vocabularies are better representations of the world than others, as opposed to being better tools for dealing with the world for one or another purpose.

To drop the idea of languages as representations, and to be thoroughly Wittgensteinian in our approach to language, would be to de-divinize the world. Only if we do that can we fully accept the argument I offered earlier – the argument that since truth is a property of sentences, since sentences are dependent for their existence upon vocabularies, and since vocabularies are made by human beings, so are truths. For as long as we think that "the world" names something we ought to respect as well as cope with, something personlike in that it has a preferred description of itself, we shall insist that any philosophical account of truth save the "intuition" that truth is "out there." This institution amounts to the vague sense that it would be *hybris* on our part to abandon the traditional language of "respect for fact" and "objectivity" – that it would be risky, and blasphemous, not to see the scientist (or the philosopher, or the poet, or *somebody*) as having a priestly function, as putting us in touch with a realm which transcends the human.

On the view I am suggesting, the claim that an "adequate" philosophical doctrine must make room for our intuitions is a reactionary slogan, one which begs the question at hand.[12] For it is essential to my view that we have no prelinguistic consciousness to which language needs to be adequate, no deep sense of how things are which it is the duty of philosophers to spell out in language. What is described as such a consciousness is simply a disposition to use the language of our ancestors, to worship the corpses of their metaphors. Unless we suffer from what Derrida calls

11 See Hans Blumenberg, *The Legitimacy of the Modern Age,* trans. Robert Wallace (Cambridge, Mass.: MIT Press, 1982).
12 For an application of this dictum to a particular case, see my discussion of the appeals to intuition found in Thomas Nagel's view of "subjectivity" and in John Searle's doctrine of "intrinsic intentionality," in "Contemporary Philosophy of Mind." For further criticisms of both, criticisms which harmonize with my own, see Daniel Dennett, "Setting Off on the Right Foot" and "Evolution, Error, and Intentionality," in Dennett, in *The Intentional Stance* (Cambridge, Mass.: MIT Press, 1987).

"Heideggerian nostalgia," we shall not think of our "intuitions" as more than platitudes, more than the habitual use of a certain repertoire of terms, more than old tools which as yet have no replacements.

I can crudely sum up the story which historians like Blumenberg tell by saying that once upon a time we felt a need to worship something which lay beyond the visible world. Beginning in the seventeenth century we tried to substitute a love of truth for a love of God, treating the world described by science as a quasi divinity. Beginning at the end of the eighteenth century we tried to substitute a love of ourselves for a love of scientific truth, a worship of our own deep spiritual or poetic nature, treated as one more quasi divinity.

The line of thought common to Blumenberg, Nietzsche, Freud, and Davidson suggests that we try to get to the point where we no longer worship *anything*, where we treat *nothing* as a quasi divinity, where we treat *everything* – our language, our conscience, our community – as a product of time and chance. To reach this point would be, in Freud's words, to "treat chance as worthy of determining our fate." In the next chapter I claim that Freud, Nietzsche, and Bloom do for our conscience what Wittgenstein and Davidson do for our language, namely, exhibit its sheer contingency.

2

The contingency of selfhood

As I was starting to write on the topic of this chapter, I came across a poem by Philip Larkin which helped me pin down what I wanted to say. Here is the last part of it:

> And once you have walked the length of your mind, what
> You command is as clear as a lading-list
> Anything else must not, for you, be thought
>   To exist.
> And what's the profit? Only that, in time
> We half-identify the blind impress
> All our behavings bear, may trace it home.
>   But to confess,
> On that green evening when our death begins,
> Just what it was, is hardly satisfying,
> Since it applied only to one man once,
>   And that man dying.

This poem discusses the fear of dying, of extinction, to which Larkin confessed in interviews. But "fear of extinction" is an unhelpful phrase. There is no such thing as fear of inexistence as such, but only fear of some concrete loss. "Death" and "nothingness" are equally resounding, equally empty terms. To say one fears either is as clumsy as Epicurus's attempt to say why one should not fear them. Epicurus said, "When I am, death is not, and when death is, I am not"; thus exchanging one vacuity for another. For the word "I" is as hollow as the word "death." To unpack such words, one has to fill in the details about the I in question, specify precisely what it is that will not be, make one's fear concrete.

Larkin's poem suggests a way of unpacking what Larkin feared. What he fears will be extinguished is his idiosyncratic lading-list, his individual sense of what was possible and important. That is what made his I different from all the other I's. To lose that difference is, I take it, what any poet – any maker, anyone who hopes to create something new – fears. Anyone who spends his life trying to formulate a novel answer to the question of what is possible and important fears the extinction of that answer.

But this does not mean simply that one fears that one's works will be lost or ignored. For that fear blends into the fear that, even if they are preserved and noticed, nobody will find anything distinctive in them. The words (or shapes, or theorems, or models of physical nature) marshaled to one's command may seem merely stock items, rearranged in routine ways. One will not have impressed one's mark on the language but, rather, will have spent one's life shoving about already coined pieces. So one will not really have had an I at all. One's creations, and one's self, will just be better or worse instances of familiar types. This is what Harold Bloom calls "the strong poet's anxiety of influence," his "horror of finding himself to be only a copy or a replica."[1]

On this reading of Larkin's poem, what would it be to have succeeded in tracing home the "blind impress" which all one's "behavings bear"? Presumably it would be to have figured out what was distinctive about oneself — the difference between one's own lading-list and other people's. If one could get this recognition down on paper (or canvas or film) — if one could find distinctive words or forms for one's own distinctiveness — then one would have *demonstrated* that one was not a copy or a replica. One would have been as strong as any poet has ever been, which means having been as strong as any human being could possibly be. For one would know exactly what it is that will die, and thus know what one has succeeded in becoming.

But the end of Larkin's poem seems to reject this Bloomian reading. There we are told that it is "hardly satisfying" to trace home one's own distinctiveness. This seems to mean that it is hardly satisfying to have become an individual — in the strong sense in which the genius is the paradigm of individuality. Larkin is affecting to despise his own vocation, on the ground that to succeed in it would merely be to have put down on paper something which "applied only to one man once / And that one dying."

I say "affecting" because I doubt that any poet could seriously think trivial his own success in tracing home the blind impress borne by all his behavings — all his previous poems. Since the example of the Romantics, since the time when, with Hegel, we began to think of self-consciousness as self-creation, no poet has seriously thought of idiosyncrasy as an objection to his work. But in this poem Larkin is pretending that blind

---

1 Harold Bloom, *The Anxiety of Influence* (Oxford University Press, 1973), p. 80. See also Bloom's claim (p. 10) that "every poet begins (however 'unconsciously') by rebelling more strongly against the fear of death than all other men and women do." I assume that Bloom would be willing to extend the reference of "poet" beyond those who write verse, and to use it in the large, generic sense in which I am using it — so that Proust and Nabokov, Newton and Darwin, Hegel and Heidegger, also fall under the term. Such people are also to be thought of as rebelling against "death" — that is, against the failure to have created — more strongly than most of us.

impresses, those particular contingencies which make each of us "I" rather than a copy or replica of somebody else, do not really matter. He is suggesting that unless one finds something common to all men at all times, not just to one man once, one cannot die satisfied. He is pretending that to be a strong poet is not enough – that he would have attained satisfaction only from being a philosopher, from finding continuities rather than exhibiting a discontinuity.[2]

I think Larkin's poem owes its interest and its strength to this reminder of the quarrel between poetry and philosophy, the tension between an effort to achieve self-creation by the recognition of contingency and an effort to achieve universality by the transcendence of contingency. The same tension has pervaded philosophy since Hegel's time,[3] and particularly since Nietzsche. The important philosophers of our own century are those who have tried to follow through on the Romantic poets by

2 "Critics, in their secret hearts, love continuities, but he who lives with continuity alone cannot be a poet" (Bloom, *Anxiety of Influence*, p. 78). The critic is, in this respect, a species of philosopher – or, more exactly, of what Heidegger and Derrida call "metaphysician." Metaphysics, Derrida says, is the search for "a centered structure . . . the concept of play as based on a fundamental ground, a play constituted on the basis of a fundamental immobility and a reassuring certitude. which is itself beyond the reach of play" (Derrida, *Writing and Difference* [Chicago: University of Chicago Press, 1978], p. 279). Metaphysicians look for continuities – overarching conditions of possibility – which provide the space within which discontinuity occurs. The secret dream of criticism is to have a pigeonhole available into which any future poet can fit; the explicit hope of pre-Kuhnian philosophers of science was to have an account of "the nature of science" which no future scientific revolution could disturb.

The most important difference between Bloom and Paul de Man (not to mention what Bloom calls the "Deconstruction Road Company") is that de Man thought philosophy had given him a sense of the necessary condition of all possible poetry – past, present, and future. I think that Bloom is right in rejecting de Man's claim that "every authentic poetic or critical act rehearses the random, meaningless act of death, for which another term is the problematic of language" (Bloom, *Agon* [Oxford University Press, 1982], p. 29). Bloom will have no truck with philosophic notions like "the problematic of language," or with abstractions like "the random, meaningless act of death." He rightly thinks that these hinder criticism, defined as the "art of knowing the hidden roads that go from poem to poem" (*Anxiety of Influence*, p. 96). Like Freud's pursuit of the hidden roads that go from the child to the adult, or from the parent to the child, such an art owes very little to the search for continuities, even the continuities posited by Freud's own metapsychology.

3 Bloom says, "If this book's argument is correct, then the covert subject of most poetry for the last three centuries has been the anxiety of influence, each poet's fear that no proper work remains for him to perform" (*Anxiety of Influence*, p. 148). I take it that Bloom would agree that this fear is common to original painters, original physicists, and original philosophers as well. In Chapter 5, I suggest that Hegel's *Phenomenology* was the book which began philosophy's period of belatedness and anxiety, the one which set the task for Nietzsche, Heidegger, and Derrida – the task of being something more than another ride on the same old dialectical seesaw. Hegel's sense of a *pattern* in philosophy was what Nietzsche called a "disadvantage of history for [the original philosopher's] life," for it suggested to Kierkegaard as well as to Nietzsche, that *now*, given Hegelian self-consciousness, there can no longer be such a thing as philosophical creativity.

breaking with Plato and seeing freedom as the recognition of contingency. These are the philosophers who try to detach Hegel's insistence on historicity from his pantheistic idealism. They accept Nietzsche's identification of the strong poet, the maker, as humanity's hero – rather than the scientist, who is traditionally pictured as a finder. More generally, they have tried to avoid anything that smacks of philosophy as contemplation, as the attempt to see life steadily and see it whole, in order to insist on the sheer contingency of individual existence.

They thus find themselves in the same sort of awkward, but interesting, position as Larkin. Larkin writes a poem about the unsatisfactoriness, compared with what pre-Nietzschean philosophers hoped to do, of doing the only thing that poets can do. Post-Nietzschean philosophers like Wittgenstein and Heidegger write philosophy in order to exhibit the universality and necessity of the individual and contingent. Both philosophers became caught up in the quarrel between philosophy and poetry which Plato began, and both ended by trying to work out honorable terms on which philosophy might surrender to poetry.

I can spell out this comparison by returning to Larkin's poem. Consider Larkin's suggestion that one might get more satisfaction out of finding a "blind impress" which applied *not* only to "one man once" but, rather, to all human beings. Think of finding such an impress as being the discovery of the universal conditions of human existence, the great continuities – the permanent, ahistorical, context of human life. This is what the priests once claimed to have done. Later the Greek philosophers, still later the empirical scientists, and later still the German idealists, made the same claim. They were going to explain to us the ultimate locus of power, the nature of reality, the conditions of the possibility of experience. They would thereby inform us what we really are, what we are compelled to be by powers not ourselves. They would exhibit the stamp which had been impressed on *all* of us. This impress would not be blind, because it would not be a matter of chance, a mere contingency. It would be necessary, essential, telic, constitutive of what it is to be a human. It would give us a goal, the only possible goal, namely, the full recognition of that very necessity, the self-consciousness of our essence.

In comparison with this universal impress, so the pre-Nietzschean philosopher's story goes, the particular contingencies of individual lives are unimportant. The mistake of the poets is to waste words on idiosyncrasies, on contingencies – to tell us about accidental appearance rather than essential reality. To admit that mere spatiotemporal location, mere contingent circumstance, mattered would be to reduce us to the level of a dying animal. To understand the context in which we necessarily live, by contrast, would be to give us a mind exactly as long as the universe itself, a lading-list which was a copy of the universe's own list. What counted as existing, as possible, or as important, for us, would be what really *is* possible, or important. Having copied this list, one could die with satisfaction, having accomplished the only task laid upon humanity, *to know the truth,* to be in touch with what is "out there." There would be nothing more to do, and thus no possible loss to be feared. Extinction would not matter, for one would have become identical with the truth, and truth, on this traditional view, is imperishable. What was extinguished would be merely idiosyncratic animality. The poets, who are not interested in truth, merely distract us from this paradigmatically human task, and thereby degrade us.

It was Nietzsche who first explicitly suggested that we drop the whole idea of "knowing the truth." His definition of truth as a "mobile army of metaphors" amounted to saying that the whole idea of "representing reality" by means of language, and thus the idea of finding a single context for all human lives, should be abandoned. His perspectivism amounted to the claim that the universe had no lading-list to be known, no determinate length. He hoped that once we realized that Plato's "true world" was just a fable, we would seek consolation, at the moment of death, not in having transcended the animal condition but in being that peculiar sort of dying animal who, by describing himself in his own terms, had created himself. More exactly, he would have created the only part of himself that mattered by constructing his own mind. To create one's mind is to create one's own language, rather than to let the length of one's mind be set by the language other human beings have left behind.[4]

But in abandoning the traditional notion of truth, Nietzsche did not abandon the idea of discovering the causes of our being what we are. He did not give up the idea that an individual might track home the blind impress all his behavings bore. He only rejected the idea that this tracking was a process of discovery. In his view, in achieving this sort of self-knowledge we are not coming to know a truth which was out there (or in here) all the time. Rather, he saw self-knowledge as self-creation. The process of coming to know oneself, confronting one's contingency, tracking one's causes home, is identical with the process of inventing a new language — that is, of thinking up some new metaphors. For any *literal* description of one's individuality, which is to say any use of an inherited language-game for this purpose, will necessarily fail. One will not have traced that idiosyncrasy home but will merely have managed to

4 My account of Nietzsche owes a great deal to Alexander Nehamas's original and penetrating *Nietzsche: Life as Literature* (Cambridge, Mass.: Harvard University Press, 1985).

see it as not idiosyncratic after all, as a specimen reiterating a type, a copy or replica of something which has already been identified. To fail as a poet – and thus, for Nietzsche, to fail as a human being – is to accept somebody else's description of oneself, to execute a previously prepared program, to write, at most, elegant variations on previously written poems. So the only way to trace home the causes of one's being as one is would be to tell a story about one's causes in a new language.

This may sound paradoxical, because we think of *causes* as discovered rather than invented. We think of telling a *causal* story as a paradigm of the *literal* use of language. Metaphor, linguistic novelty, seems out of place when one turns from simply relishing such novelty to explaining why these novelties, and not others, occurred. But remember the claim made in Chapter 1 that even in the natural sciences we occasionally get genuinely new causal stories, the sort of stories produced by what Kuhn calls "revolutionary science." Even in the sciences, metaphoric redescriptions are the mark of genius and of revolutionary leaps forward. If we follow up this Kuhnian point by thinking, with Davidson, of the literal-metaphorical distinction as the distinction between old language and new language rather than in terms of a distinction between words which latch on to the world and those which do not, the paradox vanishes. If, with Davidson, we drop the notion of language as fitting the world, we can see the point of Bloom's and Nietzsche's claim that the strong maker, the person who uses words as they have never before been used, is best able to appreciate her own contingency. For she can see, more clearly than the continuity-seeking historian, critic, or philosopher, that her *language* is as contingent as her parents or her historical epoch. She can appreciate the force of the claim that "truth is a mobile army of metaphors" because, by her own sheer strength, she has broken out of one perspective, one metaphoric, into another.

Only poets, Nietzsche suspected, can truly appreciate contingency. The rest of us are doomed to remain philosophers, to insist that there is really only one true lading-list, one true description of the human situation, one universal context of our lives. We are doomed to spend our conscious lives trying to escape from contingency rather than, like the strong poet, acknowledging and appropriating contingency. For Nietzsche, therefore, the line between the strong poet and the rest of the human race has the moral significance which Plato and Christianity attached to the distinction between the human and the animal. For although strong poets are, like all other animals, causal products of natural forces, they are products capable of telling the story of their own production in words never used before. The line between weakness and strength is thus the line between using language which is familiar and universal and producing language which, though initially unfamiliar and idiosyncratic, somehow makes tangible the blind impress all one's behavings bear. With luck – the sort of luck which makes the difference between genius and eccentricity – that language will also strike the next generation as inevitable. *Their* behavings will bear that impress.

To put the same point in another way, the Western philosophical tradition thinks of a human life as a triumph just insofar as it breaks out of the world of time, appearance, and idiosyncratic opinion into another world – into the world of enduring truth. Nietzsche, by contrast, thinks the important boundary to cross is not the one separating time from atemporal truth but rather the one which divides the old from the new. He thinks a human life triumphant just insofar as it escapes from inherited descriptions of the contingencies of its existence and finds new descriptions. This is the difference between the will to truth and the will to self-overcoming. It is the difference between thinking of redemption as making contact with something larger and more enduring than oneself and redemption as Nietzsche describes it: "recreating all 'it was' into a 'thus I willed it.' "

The drama of an individual human life, or of the history of humanity as a whole, is not one in which a preexistent goal is triumphantly reached or tragically not reached. Neither a constant external reality nor an unfailing interior source of inspiration forms a background for such dramas. Instead, to see one's life, or the life of one's community, as a dramatic narrative is to see it as a process of Nietzschean self-overcoming. The paradigm of such a narrative is the life of the genius who can say of the relevant portion of the past, "Thus I willed it," because she has found a way to describe that past which the past never knew, and thereby found a self to be which her precursors never knew was possible.

In this Nietschean view, the impulse to think, to inquire, to reweave oneself ever more thoroughly, is not wonder but terror. It is, once again, Bloom's "horror of finding oneself to be only a copy or replica." The wonder in which Aristotle believed philosophy to begin was wonder at finding oneself in a world larger, stronger, nobler than oneself. The fear in which Bloom's poets begin is the fear that one might end one's days in such a world, a world one never made, an inherited world. The hope of such a poet is that what the past tried to do to her she will succeed in doing to the past: to make the past itself, including those very causal processes which blindly impressed all her own behavings, bear *her* impress. Success in that enterprise – the enterprise of saying "Thus I willed it" to the past – is success in what Bloom calls "giving birth to oneself."

Freud's importance is that he helps us accept, and put to work, this Nietz-schean and Bloomian sense of what it is to be a full-fledged human being. Bloom has called Freud "inescapable, since more even than Proust his was the mythopoeic mind of our age, as much our theologian and our moral philosopher as he was our psychologist and our prime maker of fictions."[5] We can begin to understand Freud's role in our culture by seeing him as the moralist who helped de-divinize the self by tracking conscience home to its origin in the contingencies of our upbringing.[6]

To see Freud this way is to see him against the background of Kant. The Kantian notion of conscience divinizes the self. Once we give up, as Kant did, on the idea that scientific knowledge of hard facts is our point of contact with a power not ourselves, it is natural to do what Kant did: to turn inward, to find that point of contact in our moral consciousness – in our search for righteousness rather than our search for truth. Righteousness "deep within us" takes the place, for Kant, of empirical truth "out there." Kant was willing to let the starry heavens above be merely a *symbol* of the moral law within – an optional metaphor, drawn from the realm of the phenomenal, for the illimitableness, the sublimity, the un-conditioned character of the moral self, of that part of us which was not phenomenal, not a product of time and chance, not an effect of natural, spatiotemporal, causes.

This Kantian turn helped set the stage for the Romantic appropriation of the inwardness of the divine, but Kant himself was appalled at the Romantic attempt to make idiosyncratic poetic imagination, rather than what he called the "common moral consciouness," the center of the self. Ever since Kant's day, however, romanticism and moralism, the insis-tence on individual spontaneity and private perfection and the insistence on universally shared social responsibility, have warred with each other. Freud helps us to end this war. He de-universalizes the moral sense, making it as idiosyncratic as the poet's inventions. He thus lets us see the moral consciousness as historically conditioned, a product as much of time and chance as of political or aesthetic consciousness.

Freud ends his essay on da Vinci with a passage from which I quoted a fragment at the end of Chapter 1. He says:

5 Bloom, *Agon,* pp. 43–44. See also Harold Bloom, *Kabbalah and Criticism* (New York: Seabury Press, 1975), p. 112: "It is a curiosity . . . of much nineteenth- and twentieth-century discourse about both the nature of the human, and about ideas, that the discourse is remarkably clarified if we substitute 'poem' for 'person,' or 'poem' for 'idea.' . . . Nietzsche and Freud seem to me to be major instances of this surprising displacement."
6 I have enlarged on this claim in "Freud and Moral Reflection," in *Pragmatism's Freud,* ed. Joseph Smith and William Kerrigan (Baltimore: Johns Hopkins University Press, 1986).

If one considers chance to be unworthy of determining our fate, it is simply a relapse into the pious view of the Universe which Leonardo himself was on the way to overcoming when he wrote that the sun does not move. . . . we are all too ready to forget that in fact everything to do with our life is chance, from our origin out of the meeting of spermatozoon and ovum onwards. . . . We all still show too little respect for Nature which (in the obscure words of Leonardo which recall Hamlet's lines) "is full of countless causes ('ragioni') that never enter experience."

Every one of us human beings corresponds to one of the countless experiments in which these "ragioni" of nature force their way into experience.[7]

The commonsense Freudianism of contemporary culture makes it easy to see our conscience as such an experiment, to identify the bite of conscience with the renewal of guilt over repressed infantile sexual impulses — repressions which are the products of countless contingencies that never enter experience. It is hard nowadays to recapture how startling it must have been when Freud first began to describe conscience as an ego ideal set up by those who are "not willing to forgo the narcissistic perfection of . . . childhood."[8] If Freud had made only the large, abstract, quasi-philosophical claim that the voice of conscience is the internalized voice of parents and society, he would not have startled. That claim was suggested by Thrasymachus in Plato's *Republic,* and later developed by reductionist writers like Hobbes. What is new in Freud is the *details* he gives us about the sort of thing which goes into the formation of conscience, his explanations of why certain very concrete situations and persons excite unbearable guilt, intense anxiety, or smoldering rage. Consider, for example, the following description of the latency period:

In addition to the destruction of the Oedipus complex a regressive degradation of the libido takes place, the super-ego becomes exceptionally severe and unkind, and the ego, in obedience to the super-ego, produces strong reaction-formations in the shape of conscientiousness, pity and cleanliness. . . . But here too obsessional neurosis is only overdoing the normal method of getting rid of the Oedipus complex.[9]

This passage, and others which discuss what Freud calls "the narcissistic origin of compassion,"[10] give us a way of thinking of the sense of pity not as an identification with the common human core which we share with all

---

7 Standard Edition (S.E.), XI, 137. I owe my knowledge of this passage to William Kerrigan.
8 "On Narcissism," S.E. XIV, 94.
9 S.E. XX, 115.
10 E.g., S.E. XVII, 88.

other members of our species, but as channeled in very specific ways toward very specific sorts of people and very particular vicissitudes. He thus helps us understand how we can take endless pains to help one friend and be entirely oblivious of the greater pain of another, one whom we think we love quite as dearly. He helps explain how someone can be both a tender mother and a merciless concentration-camp guard, or be a just and temperate magistrate and also a chilly, rejecting father. By associating conscientiousness with cleanliness, and by associating both not only with obsessional neurosis but (as he does elsewhere) with the religious impulse and with the urge to construct philosophical systems, he breaks down all the traditional distinctions between the higher and the lower, the essential and the accidental, the central and the peripheral. He leaves us with a self which is a tissue of contingencies rather than an at least potentially well-ordered system of faculties.

Freud shows us why we deplore cruelty in some cases and relish it in others. He shows us why our ability to love is restricted to some very particular shapes and sizes and colors of people, things, or ideas. He shows us why our sense of guilt is aroused by certain very specific, and in theory quite minor, events, and not by others which, on any familiar moral theory, would loom much larger. Further, he gives each of us the equipment to construct our own private vocabulary of moral deliberation. For terms like "infantile" or "sadistic" or "obsessional" or "paranoid," unlike the names of vices and virtues which we inherit from the Greeks and the Christians, have very specific and very different resonances for each individual who uses them: They bring to our minds resemblances and differences between ourselves and very particular people (our parents, for example) and between the present situation and very particular situations of our past. They enable us to sketch a narrative of our own development, our idiosyncratic moral struggle, which is far more finely textured, far more custom-tailored to our individual case, than the moral vocabulary which the philosophical tradition offered us.

One can sum up this point by saying that Freud makes moral deliberation just as finely grained, just as detailed and as multiform as prudential calculation has always been. He thereby helps break down the distinction between moral guilt and practical inadvisability, thereby blurring the prudence-morality distinction. By contrast, Plato's and Kant's moral philosophies center around this distinction – as does "moral philosophy" in the sense in which it is typically understood by contemporary analytic philosophers. Kant splits us into two parts, one called "reason," which is identical in us all, and another (empirical sensation and desire), which is a matter of blind, contingent, idiosyncratic impressions. In contrast, Freud treats rationality as a mechanism which adjusts contingencies to other contingencies. But his mechanization of reason is not just more abstract philosophical reductionism, not just more "inverted Platonism." Rather than discuss rationality in the abstract, simplistic, and reductionist way in which Hobbes and Hume discuss it (a way which retains Plato's original dualisms for the sake of inverting them), Freud spends his time exhibiting the extraordinary sophistication, subtlety, and wit of our unconscious strategies. He thereby makes it possible for us to see science and poetry, genius and psychosis – and, most importantly, morality and prudence – not as products of distinct faculties but as alternative modes of adaptation.

Freud thus helps us take seriously the possibility that there is no central faculty, no central self, called "reason" – and thus to take Nietzschean pragmatism and perspectivalism seriously. Freudian moral psychology gives us a vocabulary for self-description which is radically different from Plato's, and also radically different from that side of Nietzsche which Heidegger rightly condemned as one more example of inverted Platonism – the romantic attempt to exalt the flesh over the spirit, the heart over the head, a mythical faculty called "will" over an equally mythical one called "reason."

The Platonic and Kantian idea of rationality centers around the idea that we need to bring particular actions under general principles if we are to be moral.[11] Freud suggests that we need to return to the particular – to see particular present situations and options as similar to or different from particular past actions or events. He thinks that only if we catch hold of some crucial idiosyncratic contingencies in our past shall we be able to make something worthwhile out of ourselves, to create present selves whom we can respect. He taught us to interpret what we are doing, or thinking of doing, in terms of, for example, our past reaction to particular authority-figures, or in terms of constellations of behavior which were forced upon us in infancy. He suggested that we praise ourselves by weaving idiosyncratic narratives – case histories, as it were – of our success in self-creation, our ability to break free from an idiosyncratic past. He suggests that we condemn ourselves for failure to break free of that past rather than for failure to live up to universal standards.

Another way of putting this point is that Freud gave up Plato's attempt to bring together the public and the private, the parts of the state and the parts of the soul, the search for social justice and the search for individual

---

11 For doubts about this assumption within recent analytic philosophy, see the writings of J. B. Schneewind and Annette Baier. See also Jeffrey Stout, *Ethics After Babel* (Boston: Beacon Press, 1988).

perfection. Freud gave equal respect to the appeals of moralism and romanticism, but refused either to grant one of these priority over the other or to attempt a synthesis of them. He distinguished sharply between a private ethic of self-creation and a public ethic of mutual accommodation. He persuades us that there is no bridge between them provided by universally shared beliefs or desires – beliefs or desires which belong to us qua human and which unite us to our fellow humans simply *as* human.

In Freud's account, our conscious private goals are as idiosyncratic as the unconscious obsessions and phobias from which they have branched off. Despite the efforts of such writers as Fromm and Marcuse, Freudian moral psychology cannot be used to define social goals, goals for humanity as opposed to goals for individuals. There is no way to force Freud into a Platonic mold by treating him as a moral philosopher who supplies universal criteria for goodness or rightness or true happiness. His *only* utility lies in his ability to turn us away from the universal to the concrete, from the attempt to find necessary truths, ineliminable beliefs, to the idiosyncratic contingencies of our individual pasts, to the blind impress all our behavings bear. He has provided us with a moral psychology which is compatible with Nietzsche's and Bloom's attempt to see the strong poet as the archetypal human being.

But though Freud's moral psychology is compatible with this attempt, it does not entail it. For those who share this sense of the poet as paradigmatic, Freud will seem liberating and inspiring. But suppose that, like Kant, one instead sees the unselfish, unselfconscious, unimaginative, decent, honest, dutiful person as paradigmatic. These are the people in praise of whom Kant wrote – people who, unlike Plato's philosopher, have no special acuity of mind or intellectual curiosity and who, unlike the Christian saint, are not aflame to sacrifice themselves for love of the crucified Jesus.

It was for the sake of such persons that Kant distinguished practical from pure reason, and rational religion from enthusiasm. It was for their sake that he invented the idea of a single imperative under which morality could be subsumed. For, he thought, the glory of such people is that they recognize themselves as under an unconditional obligation – an obligation which can be carried out without recourse to prudential calculation, imaginative projection, or metaphoric redescription. So Kant developed not only a novel and imaginative moral psychology but a sweeping metaphoric redescription of every facet of life and culture, precisely in order to make the intellectual world safe for such people. In his words, he denied knowledge in order to make room for faith, the faith of such people that in doing their duty they are doing all they need do, that they are paradigmatic human beings.

It has often seemed necessary to choose between Kant and Nietzsche, to make up one's mind – at least to *that* extent – about the point of being human. But Freud gives us a way of looking at human beings which helps us evade the choice. After reading Freud we shall see neither Bloom's strong poet nor Kant's dutiful fulfiller of universal obligations as paradigmatic. For Freud himself eschewed the very idea of a paradigm human being. He does not see humanity as a natural kind with an intrinsic nature, an intrinsic set of powers to be developed or left undeveloped. By breaking with both Kant's residual Platonism and Nietzsche's inverted Platonism, he lets us see both Nietzsche's superman and Kant's common moral consciousness as exemplifying two out of many forms of adaptation, two out of many strategies for coping with the contingencies of one's upbringing, of coming to terms with a blind impress. There is much to be said for both. Each has advantages and disadvantages. Decent people are often rather dull. Great wits are sure to madness near allied. Freud stands in awe before the poet, but describes him as infantile. He is bored by the merely moral man, but describes him as mature. He does not enthuse over either, nor does he ask us to choose between them. He does not think we have a faculty which can make such choices.

He does not see a need to erect a theory of human nature which will safeguard the interests of the one or the other. He sees both sorts of person as doing the best they can with the materials at their disposal, and neither as "more truly human" than the other. To abjure the notion of the "truly human" is to abjure the attempt to divinize the self as a replacement for a divinized world, the Kantian attempt I sketched at the end of Chapter 1. It is to get rid of the last citadel of necessity, the last attempt to see us as all confronting the same imperatives, the same unconditional claims. What ties Nietzsche and Freud together is this attempt – the attempt to see a blind impress as not unworthy of programming our lives or our poems.

But there is a difference between Nietzsche and Freud which my description of Freud's view of the moral man as decent but dull does not capture. Freud shows us that if we look inside the *bien-pensant* conformist, if we get him on the couch, we will find that he was only dull on the surface. For Freud, nobody is dull through and through, for there is no such thing as a dull unconscious. What makes Freud more useful and more plausible than Nietzsche is that he does not relegate the vast majority of humanity to the status of dying animals. For Freud's account of unconscious fantasy shows us how to see every human life as a poem – or, more exactly, every human life not so racked by pain as to be unable

CONTINGENCY

to learn a language nor so immersed in toil as to have no leisure in which to generate a self-description.[12] He sees every such life as an attempt to clothe itself in its own metaphors. As Philip Rieff puts it, "Freud democratized genius by giving everyone a creative unconscious."[13] The same point is made by Lionel Trilling, who said Freud "showed us that poetry is indigenous to the very constitution of the mind; he saw the mind as being, in the greater part of its tendency, exactly a poetry-making faculty."[14] Leo Bersani broadens Rieff's and Trilling's point when he says, "Psychoanalytic theory has made the notion of fantasy so richly problematic that we should no longer be able to take for granted the distinction between art and life"[15]

To say with Trilling that the mind is a poetry-making faculty may seem to return us to philosophy, and to the idea of an intrinsic human nature. Specifically, it may seem to return us to a Romantic theory of human nature in which "Imagination" plays the role which the Greeks assigned to "Reason." But it does not. "Imagination" was, for the Romantics, a link with something not ourselves, a proof that we were here as from another world. It was a faculty of expression. But what Freud takes to be shared by all of us relatively leisured language-users – all of us who have the equipment and the time for fantasy – is a faculty for creating metaphors.

In the Davidsonian account of metaphor, which I summarized in Chapter 1, when a metaphor is created it does not *express* something which previously existed, although, of course, it is *caused by* something that previously existed. For Freud, this cause is not the recollection of another world but rather some particular obsession-generating cathexis of some particular person or object or word early in life. By seeing every human being as consciously or unconsciously acting out an idiosyncratic fantasy, we can see the distinctively human, as opposed to animal, portion of each human life as the use for symbolic purposes of every particular person, object, situation, event, and word encountered in later life.

12 On the need for such a qualification, see Elaine Scarry's remarkable *The Body in Pain: The Making and Unmaking of the World* (Oxford University Press, 1985). In this book Scarry contrasts mute pain, the sort of pain which the torturer hopes to create in his victim by depriving him of language and thereby of a connection with human institutions, with the ability to share in such institutions which is given by the possession of language and leisure. Scarry points out that what the torturer really enjoys is *humiliating* his victim rather than making him scream in agony. The scream is merely one more humiliation. I develop this latter point in connection with Nabokov's and Orwell's treatments of cruelty in Chapters 7 and 8.
13 Philip Rieff, *Freud: The Mind of the Moralist* (New York: Harper & Row, 1961), p. 36.
14 Lionel Trilling, *Beyond Culture* (New York: Harcourt Brace, 1965), p. 79.
15 Leo Bersani, *Baudelaire and Freud* (Berkeley: University of California Press, 1977), p. 138.

36

This process amounts to redescribing them, thereby saying of them all, "Thus I willed it."

Seen from this angle, the intellectual (the person who uses words or visual or musical forms for this purpose) is just a special case – just somebody who does with marks and noises what other people do with their spouses and children, their fellow workers, the tools of their trade, the cash accounts of their businesses, the possessions they accumulate in their homes, the music they listen to, the sports they play or watch, or the trees they pass on their way to work. Anything from the sound of a word through the color of a leaf to the feel of a piece of skin can, as Freud showed us, serve to dramatize and crystallize a human being's sense of self-identity. For any such thing can play the role in an individual life which philosophers have thought could, or at least should, be played only by things which were universal, common to us all. It can symbolize the blind impress all our behavings bear. Any seemingly random constellation of such things can set the tone of a life. Any such constellation can set up an unconditional commandment to whose service a life may be devoted – a commandment no less unconditional because it may be intelligible to, at most, only one person.

Another way of making this point is to say that the social process of literalizing a metaphor is duplicated in the fantasy life of an individual. We call something "fantasy" rather than "poetry" or "philosophy" when it revolves around metaphors which do not catch on with other people – that is, around ways of speaking or acting which the rest of us cannot find a use for. But Freud shows us how something which seems pointless or ridiculous or vile to society can become the crucial element in the individual's sense of who she is, her own way of tracing home the blind impress all her behavings bear. Conversely, when some private obsession produces a metaphor which we *can* find a use for, we speak of genius rather than of eccentricity or perversity. The difference between genius and fantasy is not the difference between impresses which lock on to something universal, some antecedent reality out there in the world or deep within the self, and those which do not. Rather, it is the difference between idiosyncrasies which just happen to catch on with other people – happen because of the contingencies of some historical situation, some particular need which a given community happens to have at a given time.

To sum up, poetic, artistic, philosophical, scientific, or political progress results from the accidental coincidence of a private obsession with a public need. Strong poetry, commonsense morality, revolutionary morality, normal science, revolutionary science, and the sort of fantasy which is intelligible to only one person, are all, from a Freudian point of view, different ways of dealing with blind impresses – or, more precisely, ways of dealing with different blind impresses: impresses which may be unique to an individual or common to the members of some historically conditioned community. None of these strategies is privileged over others in the sense of expressing human nature better. No such strategy is more or less human than any other, any more than the pen is more truly a tool than the butcher's knife, or the hybridized orchid less a flower than the wild rose.

To appreciate Freud's point would be to overcome what William James called "a certain blindness in human beings." James's example of this blindness was his own reaction, during a trip through the Appalachian Mountains, to a clearing in which the forest had been hacked down and replaced with a muddy garden, a log cabin, and some pigpens. As James says, "The forest had been destroyed; and what had 'improved' it out of existence was hideous, a sort of ulcer, without a single element of artificial grace to make up for the loss of Nature's beauty." But, James continues, when a farmer comes out of the cabin and tells him that "we ain't happy here unless we're getting one of those coves under cultivation," he realizes that

I had been losing the whole inward significance of the situation. Because to me the clearings spoke of naught but denudation, I thought that to those whose sturdy arms and obedient axes had made them they could tell no other story. But when *they* looked on the hideous stumps, what they thought of was personal victory. . . . In short, the clearing which to me was a mere ugly picture on the retina, was to them a symbol redolent with moral memories and sang a very paean of duty, struggle and success.[16]

I had been as blind to the peculiar ideality of their conditions as they certainly would also have been to the ideality of mine, had they had a peep at my strange indoor academic ways of life at Cambridge.

I take Freud to have spelled out James's point in more detail, helping us overcome particularly intractable cases of blindness by letting us see the "peculiar ideality" of events which exemplify, for example, sexual perversion, extreme cruelty, ludicrous obsession, and manic delusion. He let us see each of these as the private poem of the pervert, the sadist, or the lunatic: each as richly textured and "redolent of moral memories" as our own life. He lets us see what moral philosophy describes as extreme, inhuman, and unnatural, as continuous with our own activity. But, and

16 "On a Certain Blindness in Human Beings," in James, *Talks to Teachers on Psychology,* eds. Frederick Burkhardt and Fredson Bowers (Cambridge, Mass.: Harvard University Press, 1983), p. 134.

this is the crucial point, he does not do so in the traditional philosophical, reductionist way. He does not tell us that art is *really* sublimation or philosophical system-building *merely* paranoia, or religion *merely* a confused memory of the fierce father. He is not saying that human life is *merely* a continuous rechanneling of libidinal energy. He is not interested in invoking a reality-appearance distinction, in saying that anything is "merely" or "really" something quite different. He just wants to give us one more redescription of things to be filed alongside all the others, one more vocabulary, one more set of metaphors which he thinks have a chance of being used and thereby literalized.

Insofar as one can attribute philosophical views to Freud, one can say that he is as much a pragmatist as James and as much a perspectivalist as Nietzsche – or, one might also say, as much a modernist as Proust.[17] For it somehow became possible, toward the end of the nineteenth century, to take the activity of redescription more lightly than it had ever been taken before. It became possible to juggle several descriptions of the same event without asking which one was right – to see redescription as a tool rather than a claim to have discovered essence. It thereby became possible to see a new vocabulary not as something which was supposed to replace all other vocabularies, something which claimed to represent reality, but simply as one more vocabulary, one more human project, one person's chosen metaphoric. It is unlikely that Freud's metaphors could have been picked up, used, and literalized at any earlier period. But, conversely, it is unlikely that without Freud's metaphors we should have been able to assimilate Nietzsche's, James's, Wittgenstein's, or Heidegger's as easily as we have, or to have read Proust with the relish we did. All the figures of this period play into each other's hands. They feed each other lines. Their metaphors rejoice in one another's company. This is the sort of phenomenon it is tempting to describe in terms of the march of the World-Spirit toward clearer self-consciousness, or as the length of man's mind gradually coming to match that of the universe. But any such description would betray the spirit of playfulness and irony which links the figures I have been describing.

This playfulness is the product of their shared ability to appreciate the power of redescribing, the power of language to make new and different things possible and important – an appreciation which becomes possible only when one's aim becomes an expanding repertoire of alternative

---

17 See Bloom, *Agon*, p. 23: ". . . by 'a literary culture' I do mean Western society now, since it has no authentic religion and no authentic philosophy, and will never acquire them again, and because psychoanalysis, its pragmatic religion and philosophy, is just a fragment of literary culture, so that in time we will speak alternatively of Freudianism *or* Proustianism." I discuss Proust's role as moral exemplar in Chapter 5.

descriptions rather than The One Right Description. Such a shift in aim is possible only to the extent that both the world and the self have been de-divinized. To say that both are de-divinized is to say that one no longer thinks of either as speaking to us, as having a language of its own, as a rival poet. Neither are quasi persons, neither wants to be expressed or represented in a certain way.

Both, however, have power over us – for example, the power to kill us. The world can blindly and inarticulately crush us; mute despair, intense mental pain, can cause us to blot ourselves out. But that sort of power is not the sort we can appropriate by adopting and then transforming its language, thereby becoming identical with the threatening power and subsuming it under our own more powerful selves. This latter strategy is appropriate only for coping with other persons – for example, with parents, gods, and poetic precursors. For our relation to the world, to brute power and to naked pain, is not the sort of relation we have to persons. Faced with the nonhuman, the nonlinguistic, we no longer have an ability to overcome contingency and pain by appropriation and transformation, but only the ability to *recognize* contingency and pain. The final victory of poetry in its ancient quarrel with philosophy – the final victory of metaphors of self-creation over metaphors of discovery – would consist in our becoming reconciled to the thought that this is the only sort of power over the world which we can hope to have. For that would be the final abjuration of the notion that truth, and not just power and pain, is to be found "out there."

It is tempting to suggest that in a culture in which poetry had publicly and explicitly triumphed over philosophy, a culture in which recognition of contingency rather than of necessity was the accepted definition of freedom, Larkin's poem would fall flat. There would be no pathos in finitude. But there probably cannot be such a culture. Such pathos is probably ineliminable. It is as hard to imagine a culture dominated by exuberant Nietzschean playfulness as to imagine the reign of the philosopher-kings, or the withering away of the state. It is equally hard to imagine a human life which felt itself complete, a human being who dies happy because all that he or she ever wanted has been attained.

This is true even for Bloom's strong poet. Even if we drop the philosophical ideal of seeing ourselves steadily and whole against a permanent backdrop of "literal" unchangeable fact, and substitute the ideal of seeing ourselves in our own terms, of redemption through saying to the past, "Thus I willed it," it will remain true that this willing will always be a project rather than a result, a project which life does not last long enough to complete.

The strong poet's fear of death as the fear of incompletion is a function of the fact that no project of redescribing the world and the past, no project of self-creation through imposition of one's own idiosyncratic metaphoric, can avoid being marginal and parasitic. Metaphors are unfamiliar uses of old words, but such uses are possible only against the background of other old words being used in old familiar ways. A language which was "all metaphor" would be a language which had no use, hence not a language but just babble. For even if we agree that languages are not media of representation or expression, they will remain media of communication, tools for social interaction, ways of tying oneself up with other human beings.

This needed corrective to Nietzsche's attempt to divinize the poet, this dependence of even the strongest poet on others, is summed up by Bloom as follows:

> The sad truth is that poems *don't have* presence, unity, form or meaning. . . .
> What then does a poem possess or create? Alas, a poem *has* nothing, and *creates* nothing. Its presence is a promise, part of the substance of things hoped for, the evidence of things not seen. Its unity is in the good will of the reader. . . . its meaning is just that there is, or rather *was,* another poem.[18]

In this passage Bloom de-divinizes the poem, and thereby the poet, in the same way in which Nietzsche de-divinized truth and in which Freud de-divinized conscience. He does for romanticism what Freud did for moralism. The strategy is the same in all these cases: It is to substitute a tissue of contingent relations, a web which stretches backward and forward through past and future time, for a formed, unified, present, self-contained substance, something capable of being seen steadily and whole. Bloom reminds us that just as even the strongest poet is parasitic on her precursors, just as even she can give birth only to a small part of herself, so she is dependent on the kindness of all those strangers out there in the future.

This amounts to a reminder of Wittgenstein's point that there are no private languages – his argument that you cannot give meaning to a word or a poem by confronting it with a nonlinguistic meaning, something other than a bunch of already used words or a bunch of already written poems.[19] Every poem, to paraphrase Wittgenstein, presupposes a lot of

---

18 Bloom, *Kabbalah and Criticism,* p. 122.
19 "Just as we can never embrace (sexually or otherwise) a single person, but embrace the whole of her or his family romance, so we can never read a poet without reading the whole of his or her family romance as poet. The issue is reduction and how best to avoid it. Rhetorical, Aristotelian, phenomenological, and structuralist criticisms all

stage-setting in the culture, for the same reason that every sparkling metaphor requires a lot of stodgy literal talk to serve as its foil. Shifting from the written poem to the life-as-poem, one may say that there can be no fully Nietzschean lives, lives which are pure action rather than reaction – no lives which are not largely parasitical on an un-redescribed past and dependent on the charity of as yet unborn generations. There is no stronger claim even the strongest poet can make than the one Keats made – that he "would be among the English poets," construing "among them" in a Bloomian way as "in the midst of them," future poets living out of Keats's pockets as he lived out of those of his precursors. Analogously, there is no stronger claim which even the superman can make than that his differences from the past, inevitably minor and marginal as they are, will nevertheless be carried over into the future – that his metaphoric redescriptions of small parts of the past will be among the future's stock of literal truths.

To sum up, I suggest that the best way to understand the pathos of finitude which Larkin invokes is to interpret it not as the failure to achieve what philosophy hoped to achieve – something nonidiosyncratic, atemporal, and universal – but as the realization that at a certain point one has to trust to the good will of those who will live other lives and write other poems. Nabokov built his best book, *Pale Fire,* around the phrase "Man's life as commentary to abstruse unfinished poem." That phrase serves both as a summary of Freud's claim that every human life is the working out of a sophisticated idiosyncratic fantasy, and as a reminder that no such working out gets completed before death interrupts. It cannot get completed because there is nothing to complete,

reduce, whether to images, ideas, given things, or phonemes. Moral and other blatant philosophical or psychological criticisms all reduce to rival conceptualizations. We reduce – if at all – to another poem. *The meaning of a poem can only be another poem"* (Bloom, *The Anxiety of Influence,* p. 94; italics added). See also p. 70, and compare p. 43: "Let us give up the failed enterprise of seeking to 'understand' any single poem as an entity in itself. Let us pursue instead the quest of learning to read any poem as its poet's deliberate misinterpretation, *as a poet,* of a precursor poem or of poetry in general."

There is an analogy between Bloom's antireductionism and Wittgenstein's, Davidson's and Derrida's willingness to let meaning consist in relation to other texts rather than in a relation to something outside the text. The idea of a private language, like Sellars's Myth of the Given, stems from the hope that words might get meaning without relying on other words. This hope, in turn, stems from the larger hope, diagnosed by Sartre, of becoming a self-sufficient *être-en-soi.* Sartre's description ("Portrait of the Anti-Semite," in *Existentialism from Dostoevsky to Sartre,* ed. Walter Kaufmann [New York: New American Library, 1975], p. 345) of the anti-Semite as "the man who wants to be pitiless stone, furious torrent, devastating lightning – in short, everything but a man" – is a criticism of Zarathustra, of what Bloom calls "reductionist" criticism, and of what Heidegger and Derrida call "metaphysics."

there is only a web of relations to be rewoven, a web which time lengthens every day.

But if we avoid Nietzsche's inverted Platonism — his suggestion that a life of self-creation can be as complete and as autonomous as Plato thought a life of contemplation might be — then we shall be content to think of any human life as the always incomplete, yet sometimes heroic, reweaving of such a web. We shall see the conscious need of the strong poet to *demonstrate* that he is not a copy or replica as merely a special form of an unconscious need everyone has: the need to come to terms with the blind impress which chance has given him, to make a self for himself by redescribing that impress in terms which are, if only marginally, his own.

3

The contingency of a liberal community

Anyone who says, as I did in Chapter 1, that truth is not "out there" will be suspected of relativism and irrationalism. Anyone who casts doubt on the distinction between morality and prudence, as I did in Chapter 2, will be suspected of immorality. To fend off such suspicions, I need to argue that the distinctions between absolutism and relativism, between rationality and irrationality, and between morality and expediency are obsolete and clumsy tools – remnants of a vocabulary we should try to replace. But "argument" is not the right word. For on my account of intellectual progress as the literalization of selected metaphors, rebutting objections to one's redescriptions of some things will be largely a matter of redescribing other things, trying to outflank the objections by enlarging the scope of one's favorite metaphors. So my strategy will be to try to make the vocabulary in which these objections are phrased look bad, thereby changing the subject, rather than granting the objector his choice of weapons and terrain by meeting his criticisms head-on.

In this chapter I shall claim that the institutions and culture of liberal society would be better served by a vocabulary of moral and political reflection which avoids the distinctions I have listed than by a vocabulary which preserves them. I shall try to show that the vocabulary of Enlightenment rationalism, although it was essential to the beginnings of liberal democracy, has become an impediment to the preservation and progress of democratic societies. I shall claim that the vocabulary I adumbrated in the first two chapters, one which revolves around notions of metaphor and self-creation rather than around notions of truth, rationality, and moral obligation, is better suited for this purpose.

I am not, however, saying that the Davidsonian-Wittgensteinian account of language and the Nietzschean-Freudian account of conscience and selfhood which I have sketched provide "philosophical foundations of democracy." For the notion of a "philosophical foundation" goes when the vocabulary of Enlightenment rationalism goes. These accounts do not ground democracy, but they do permit its practices and its goals to be redescribed. In what follows I shall be trying to reformulate the hopes of liberal society in a nonrationalist and nonuniveralist way – one which furthers their realization better than older descriptions of them

44

did. But to offer a redescription of our current institutions and practices is not to offer a defense of them against their enemies; it is more like refurnishing a house than like propping it up or placing barricades around it.

The difference between a search for foundations and an attempt at redescription is emblematic of the difference between the culture of liberalism and older forms of cultural life. For in its ideal form, the culture of liberalism would be one which was enlightened, secular, through and through. It would be one in which no trace of divinity remained, either in the form of a divinized world or a divinized self. Such a culture would have no room for the notion that there are nonhuman forces to which human beings should be responsible. It would drop, or drastically reinterpret, not only the idea of holiness but those of "devotion to truth" and of "fulfillment of the deepest needs of the spirit." The process of de-divinization which I described in the previous two chapters would, ideally, culminate in our no longer being able to see any use for the notion that finite, mortal, contingently existing human beings might derive the meanings of their lives from anything except other finite, mortal, contingently existing human beings. In such a culture, warnings of "relativism," queries whether social institutions had become increasingly "rational" in modern times, and doubts about whether the aims of liberal society were "objective moral values" would seem merely quaint.

To give some initial plausibility to my claim that my view is well adapted to a liberal polity, let me note some parallels between it and Isaiah Berlin's defense of "negative liberty" against telic conceptions of human perfection. In his *Two Concepts of Liberty,* Berlin says, as I did in Chapter 1, that we need to give up the jigsaw puzzle approach to vocabularies, practices, and values. In Berlin's words, we need to give up "the conviction that all the positive values in which men have believed must, in the end, be compatible, and perhaps even entail each other."[1] My emphasis on Freud's claim that we should think of ourselves as just one more among Nature's experiments, not as the culmination of Nature's design, echoes Berlin's use of J. S. Mill's phrase "experiments in living" (as well as echoing Jefferson's and Dewey's use of the term "experiment" to describe American democracy). In my second chapter I inveighed against the Platonic-Kantian attempt to do what Berlin called "splitting [our] personality into two: the transcendent, dominant controller and the em-

---

1 Isaiah Berlin, *Four Essays on Liberty* (Oxford University Press, 1969), p. 167.

pirical bundle of desires and passions to be disciplined and brought to heel."[2]

Berlin ended his essay by quoting Joseph Schumpeter, who said, "To realise the relative validity of one's convictions and yet stand for them unflinchingly, is what distinguishes a civilized man from a barbarian." Berlin comments, "To demand more than this is perhaps a deep and incurable metaphysical need; but to allow it to determine one's practice is a symptom of an equally deep, and more dangerous, moral and political immaturity."[3] In the jargon I have been developing, Schumpeter's claim that this is the mark of the civilized person translates into the claim that the liberal societies of our century have produced more and more people who are able to recognize the contingency of the vocabulary in which they state their highest hopes – the contingency of their own consciences – and yet have remained faithful to those consciences. Figures like Nietzsche, William James, Freud, Proust, and Wittgenstein illustrate what I have called "freedom as the recognition of contingency." In this chapter I shall claim that such recognition is the chief virtue of the members of a liberal society, and that the culture of such a society should aim at curing us of our "deep metaphysical need."

To show how the charge of relativism looks from my point of view, I shall discuss a comment on Berlin's essay by an acute contemporary critic of the liberal tradition, Michael Sandel. Sandel says that Berlin "comes perilously close to foundering on the relativist predicament." He asks:

If one's convictions are only relatively valid, why stand for them unflinchingly? In a tragically configured moral universe, such as Berlin assumes, is the ideal of freedom any less subject than competing ideals to the ultimate incommensurability of values? If so, in what can its privileged status consist? And if freedom has no morally privileged status, if it is just one value among many, then what can be said for liberalism?[4]

In posing these questions, Sandel is taking the vocabulary of Enlightenment rationalism for granted. Further, he is taking advantage of the fact

---

2   Ibid., p. 134.
3   Ibid., p. 172.
4   "Introduction" to Michael Sandel, ed., *Liberalism and its Critics* (New York: New York University Press, 1984), p. 8. These comments represent Sandel's account of the standard objection to Berlin rather than his own attitude. Elsewhere I have discussed Sandel's own views in some detail, and attempted to rebut some of the objections to Rawls which he formulated in his *Liberalism and the Limits of Justice* (Cambridge University Press, 1982). See my "The Priority of Democracy to Philosophy," in *The Virginia Statute for Religious Freedom*, ed. Merrill D. Peterson and Robert C. Vaughan (Cambridge University Press, 1988).

that Schumpeter and Berlin themselves make use of this vocabulary, and attempting thereby to show that their view is incoherent. Going over Sandel's questions in some detail may help make clear what sort of view people must hold who find the terms "relativism" and "morally privileged" useful. It may thus help show why it would be better to avoid using the term "only relatively valid" to characterize the state of mind of the figures whom Schumpeter, Berlin, and I wish to praise.

To say that convictions are only "relatively valid" might seem to mean that they can only be justified to people who hold certain other beliefs – not to anyone and everyone. But if this were what was meant, the term would have no contrastive force, for there would be no interesting statements which were *absolutely* valid. Absolute validity would be confined to everyday platitudes, elementary mathematical truths, and the like: the sort of beliefs nobody wants to argue about because they are neither controversial nor central to anyone's sense of who she is or what she lives for. All beliefs which *are* central to a person's self-image are so because their presence or absence serves as a criterion for dividing good people from bad people, the sort of person one wants to be from the sort one does not want to be. A conviction which can be justified to *anyone* is of little interest. "Unflinching courage" will not be required to sustain such a conviction.

So we must construe the term "only relatively valid beliefs" as contrasting with statements capable of being justified to all those who are uncorrupted – that is, to all those in whom reason, viewed as a built-in truth-seeking faculty, or conscience, viewed as a built-in righteousness detector, is powerful enough to overcome evil passions, vulgar superstitions, and base prejudices. The notion of "absolute validity" does not make sense except on the assumption of a self which divides fairly neatly into the part it shares with the divine and the part it shares with the animals. But if we accept this opposition between reason and passion, or reason and will, we liberals will be begging the question against ourselves. It behooves those of us who agree with Freud and Berlin that we should *not* split persons up into reason and passion to drop, or at least to restrict the use of, the traditional distinction between "rational conviction" and "conviction brought about by reasons rather than causes."

The best way of restricting its use is to limit the opposition between rational and irrational forms of persuasion to the interior of a language game, rather than to try to apply it to interesting and important shifts in linguistic behavior. Such a restricted notion of rationality is all we can allow ourselves if we accept the central claim of Chapter 1: that what matters in the end are changes in the vocabulary rather than changes in belief, changes in truth-value candidates rather than assignments of truth-value. Within a language game, within a set of agreements about what is possible and important, we can usefully distinguish reasons for belief from causes for belief which are not reasons. We do this by starting with such obvious differences as that between Socratic dialogue and hypnotic suggestion. We then try to firm up the distinction by dealing with messier cases: brainwashing, media hype, and what Marxists call "false consciousness." There is, to be sure, no neat way to draw the line between persuasion and force, and therefore no neat way to draw a line between a cause of changed belief which was also a reason and one which was a "mere" cause. But the distinction is no fuzzier than most.

However, once we raise the question of how we get from one vocabulary to another, from one dominant metaphoric to another, the distinction between reasons and causes begins to lose its utility. Those who speak the old language and have no wish to change, those who regard it as a hallmark of rationality or morality to speak just that language, will regard as altogether *ir*rational the appeal of the new metaphors – the new language game which the radicals, the youth, or the avant-garde are playing. The popularity of the new ways of speaking will be viewed as a matter of "fashion" or "the need to rebel" or "decadence." The question of why people speak this way will be treated as beneath the level of conversation – a matter to be turned over to psychologists or, if necessary, the police. Conversely, from the point of view of those who are trying to use the new language, to literalize the new metaphors, those who cling to the old language will be viewed as irrational – as victims of passion, prejudice, superstition, the dead hand of the past, and so on. The philosophers on either side can be counted on to support these opposing invocations of the reason-cause distinction by developing a moral psychology, or an epistemology, or a philosophy of language, which will put those on the other side in a bad light.

To accept the claim that there is no standpoint outside the particular historically conditioned and temporary vocabulary we are presently using from which to judge this vocabulary is to give up on the idea that there can be reasons for using languages as well as reasons within languages for believing statements. This amounts to giving up the idea that intellectual or political progress is rational, in any sense of "rational" which is neutral between vocabularies. But because it seems pointless to say that all the great moral and intellectual advances of European history – Christianity, Galilean science, the Enlightenment, Romanticism, and so on – were fortunate falls into temporary irrationality, the moral to be drawn is that the rational-irrational distinction is less useful than it once appeared. Once we realize that progress, for the community as for the individual, is a matter of using new words as well as of arguing from premises phrased in old words, we realize that a critical vocabulary which revolves around notions like "rational," "criteria," "argument" and "foundation" and "absolute" is badly suited to describe the relation between the old and the new.

At the conclusion of an essay on Freud's account of irrationality, Davidson notes that once we give up on the notion of "absolute criteria of rationality," and so use the term "rational" to mean something like "internal coherence," then, if we do not limit the range of this term's application, we shall be forced to call "irrational" many things we wish to praise. In particular we shall have to describe as "irrational" what Davidson calls "a form of self-criticism and reform which we hold in high esteem, and that has even been thought to be the very essence of rationality and the source of freedom." Davidson makes the point as follows:

> What I have in mind is a special kind of second-order desire or value, and the actions it can touch off. This happens when a person forms a positive or negative judgment of some of his own desires, and he acts to change these desires. From the point of view of the changed desire, there is no reason for the change – the reason comes from an independent source, and is based on further, and partly contrary, considerations. The agent has reasons for changing his own habits and character, but those reasons come from a domain of values necessarily extrinsic to the contents of the views or values to undergo change. The cause of the change, if it comes, can therefore not be a reason for what it causes. A theory that could not explain irrationality would be one that also could not explain our salutary efforts, and occasional successes, at self-criticism and self-improvement.[5]

Davidson would, of course, be wrong if self-criticism and self-improvement always take place within a framework of nontrivial highest-possible-order desires, those of the *true* self, the desires which are central to our humanity. For then these highest-level desires would mediate and rationalize the contest between first- and second-level desires. But Davidson is assuming – rightly, I think – that the only candidates for such highest-level desires are so abstract and empty as to have no mediating powers: They are typified by "I wish to be good," "I wish to be rational," and "I wish to know the truth." Because what will *count* as "good" or "rational" or "true" will be determined by the contest between the first- and second-level desires, wistful top-level protestations of goodwill are impotent to intervene in that contest.

If Davidson is right, then the assumptions usually invoked against Berlin and Schumpeter are wrong. We shall not be able to assume that

---

5 Donald Davidson, "Paradoxes of Irrationality," in *Philosophical Essays on Freud*, ed. Richard Wollheim and James Hopkins (Cambridge University Press, 1982), p. 305.

there is a largest-possible framework within which one can ask, for example, "If freedom has no morally privileged status, if it is just one value among many, then what can be said for liberalism?" We cannot assume that liberals ought to be able to rise above the contingencies of history and see the kind of individual freedom which the modern liberal state offers its citizens as just one more value. Nor can we assume that the rational thing to do is to place such freedom alongside other candidates (e.g., the sense of national purpose which the Nazis briefly offered the Germans, or the sense of conformity to the will of God which inspired the Wars of Religion) and then use "reason" to scrutinize these various candidates and discover which, if any, are "morally privileged." Only the assumption that there is some such standpoint to which we might rise gives sense to the question, "If one's convictions are only relatively valid, why stand for them unflinchingly?"

Conversely, neither Schumpeter's phrase "relative validity" nor the notion of a "relativist predicament" will seem in point if one grants Davidson's claim that new metaphors are causes, but not reasons, for changes of belief, and Hesse's claim that it is new metaphors which have made intellectual progress possible. If one grants these claims, there is no such thing as the "relativist predicament," just as for someone who thinks that there is no God there will be no such thing as blasphemy. For there will be no higher standpoint to which we are responsible and against whose precepts we might offend. There will be no such activity as scrutinizing competing values in order to see which are morally privileged. For there will be no way to rise above the language, culture, institutions, and practices one has adopted and view all these as on a par with all the others. As Davidson puts it, "speaking a language . . . is not a trait a man can lose while retaining the power of thought. So there is no chance that someone can take up a vantage point for comparing conceptual schemes by temporarily shedding his own."[6] Or, to put the point in Heidegger's way, "language speaks man," languages change in the course of history, and so human beings cannot escape their historicity. The most they can do is to manipulate the tensions within their own epoch in order to produce the beginnings of the next epoch.

But, of course, if the presuppositions of Sandel's questions are right, then Davidson and Heidegger are wrong. Davidsonian and Wittgensteinian philosophy of language — the account of language as a historical contingency rather than as a medium which is gradually taking on the true shape of the true world or the true self — will beg the question. If we accept Sandel's questions, then we shall ask instead for a philosophy of

6 Davidson, *Inquiries into Truth and Interpretation*, p. 185.

language, an epistemology, and a moral psychology which will safeguard the interests of reason, preserve a morality-prudence distinction, and thus guarantee that Sandel's questions are in point. We shall want a different way of seeing language, one which treats it as a medium in which to find truth which is out there in the world (or, at least, deep within the self, at the place where we find the permanent, ahistorical, highest-level desires which ajudicate lower-level conflicts). We shall want to refurbish the subject-object and scheme-content models of inquiry – the models which Davidson and Heidegger describe as obsolete.

Is there a way to resolve this standoff between the traditional view that it is always in point to ask "How do you know?" and the view that sometimes all we can ask is "Why do you talk that way?" Philosophy, as a discipline, makes itself ridiculous when it steps forward at such junctures and says that it will find neutral ground on which to adjudicate the issue. It is not as if the philosophers had succeeded in finding some neutral ground on which to stand. It would be better for philosophers to admit there is no *one* way to break such standoffs, no single place to which it is appropriate to step back. There are, instead, as many ways of breaking the standoff as there are topics of conversation. One can come at the issue by way of different paradigms of humanity – the contemplator as opposed to the poet, or the pious person as opposed to the person who accepts chance as worthy of determining her fate. Or one can come at it from the point of view of an ethics of kindness, and ask whether cruelty and injustice will be diminished if we all stopped worrying about "absolute validity" or whether, on the contrary, only such worries keep our characters firm enough to defend unflinchingly the weak against the strong. Or one can – fruitlessly, in my view – come at it by way of anthropology and the question of whether there are "cultural universals," or by way of psychology and the question of whether there are psychological universals. Because of this indefinite plurality of standpoints, this vast number of ways of coming at the issue sideways and trying to outflank one's opponent, there are never, in practice, any standoffs.

We would only have a real and practical standoff, as opposed to an artificial and theoretical one, if certain topics and certain language games were taboo – if there were general agreement within a society that certain questions were *always* in point, that certain questions were prior to certain others, that there was a fixed order of discussion, and that flanking movements were not permitted. That would be just the sort of society which liberals are trying to avoid – one in which "logic" ruled and "rhetoric" was outlawed. It is central to the idea of a liberal society that, in respect to words as opposed to deeds, persuasion as opposed to force, anything goes. This openmindedness should not be fostered because, as Scripture teaches, Truth is great and will prevail, nor because, as Milton suggests, Truth will always win in a free and open encounter. It should be fostered for its own sake. *A liberal society is one which is content to call "true" whatever the upshot of such encounters turns out to be.* That is why a liberal society is badly served by an attempt to supply it with "philosophical foundations." For the attempt to supply such foundations presupposes a natural order of topics and arguments which is prior to, and overrides the results of, encounters between old and new vocabularies.

This last point permits me to turn to the larger claim I put forward earlier: the claim that liberal culture needs an improved self-description rather than a set of foundations. The idea that it ought to have foundations was a result of Enlightenment scientism, which was in turn a survival of the religious need to have human projects underwritten by a nonhuman authority. It was natural for liberal political thought in the eighteenth century to try to associate itself with the most promising cultural development of the time, the natural sciences. But unfortunately the Enlightenment wove much of its political rhetoric around a picture of the scientist as a sort of priest, someone who achieved contact with nonhuman truth by being "logical," "methodical," and "objective."[7] This was a useful tactic in its day, but it is less useful nowadays. For, in the first place, the sciences are no longer the most interesting or promising or exciting area of culture. In the second place, historians of science have made clear how little this picture of the scientist has to do with actual scientific achievement, how pointless it is to try to isolate something called "the scientific method." Although the sciences have burgeoned a thousandfold since the end of the eighteenth century, and have thereby made possible the realization of political goals which could never have been realized without them, they have nevertheless receded into the background of cultural life. This recession is due largely to the increasing difficulty of mastering the various languages in which the various sciences are conducted. It is not something to be deplored but, rather, something to be coped with. We can do so by switching attention to the areas which *are* at the forefront of culture, those which excite the imagination of the young, namely, art and utopian politics.

I said at the beginning of Chapter 1 that the French Revolution and the Romantic movement inaugurated an era in which we gradually came to

7 For more on this point, see my "Science as Solidarity," in *The Rhetoric of the Human Sciences,* ed. John S. Nelson et al. (Madison: University of Wisconsin Press, 1987), pp. 38–52, and "Pragmatism Without Method," in *Sidney Hook: Philosopher of Democracy and Humanism,* ed. Paul Kurtz (Buffalo: Prometheus Books, 1983), pp. 259–273.

appreciate the historical role of linguistic innovation. This appreciation is summed up in the vague, misleading, but pregnant and inspiring thought that truth is made rather than found. I also said that literature and politics are the spheres to which contemporary intellectuals look when they worry about ends rather than about means. I can now add the corollary that these are the areas to which we should look for the charter of a liberal society. We need a redescription of liberalism as the hope that culture as a whole can be "poeticized" rather than as the Enlightenment hope that it can be "rationalized" or "scientized." That is, we need to substitute the hope that chances for fulfillment of idiosyncratic fantasies will be equalized for the hope that everyone will replace "passion" or fantasy with "reason."

In my view, an ideally liberal polity would be one whose culture hero is Bloom's "strong poet" rather than the warrior, the priest, the sage, or the truth-seeking, "logical," "objective" scientist. Such a culture would slough off the Enlightenment vocabulary which enshrines the presuppositions of Sandel's questions to Berlin. It would no longer be haunted by specters called "relativism" and "irrationalism." Such a culture would not assume that a form of cultural life is no stronger than its philosophical foundations. Instead, it would drop the idea of such foundations. It would regard the justification of liberal society simply as a matter of historical comparison with other attempts at social organization – those of the past and those envisaged by utopians.

To think such a justification sufficient would be to draw the consequences from Wittgenstein's insistence that vocabularies – all vocabularies, even those which contain the words which we take most seriously, the ones most essential to our self-descriptions – are human creations, tools for the creation of such other human artifacts as poems, utopian societies, scientific theories, and future generations. Indeed, it would be to build the rhetoric of liberalism around this thought. This would mean giving up the idea that liberalism could be justified, and Nazi or Marxist enemies of liberalism refuted, by driving the latter up against an argumentative wall – forcing them to admit that liberal freedom has a "moral privilege" which their own values lacked. From the point of view I have been commending, any attempt to drive one's opponent up against a wall in this way fails when the wall against which he is driven comes to be seen as one more vocabulary, one more way of describing things. The wall then turns out to be a painted backdrop, one more work of man, one more bit of cultural stage-setting. A poeticized culture would be one which would not insist we find the real wall behind the painted ones, the real touchstones of truth as opposed to touchstones which are merely cultural artifacts. It would be a culture which, precisely by appreciating that *all* touchstones are such artifacts, would take as its goal the creation of ever more various and multicolored artifacts.

To sum up, the moral I want to draw from my discussion of the claim that Berlin's position is "relativistic" is that this charge should not be answered, but rather evaded. We should learn to brush aside questions like "How do you *know* that freedom is the chief goal of social organization?" in the same way as we brush aside questions like "How do you *know* that Jones is worthy of your friendship?" or "How do you *know* that Yeats is an important poet, Hegel an important philosopher, and Galileo an important scientist?" We should see allegiance to social institutions as no more matters for justification by reference to familiar, commonly accepted premises – but also as no more arbitrary – than choices of friends or heroes.[8] Such choices are not made by reference to criteria. They cannot be preceded by presuppositionless critical reflection, conducted in no particular language and outside of any particular historical context.

When I say "we should do this or that "we cannot" do that, I am not, of course, speaking from a neutral standpoint. I am speaking from Berlin's side of the argument, trying to serve as an underlaborer to Berlin by clearing away some of the philosophical underbrush. I am no more neutral, and philosophy can no more be neutral, on political matters of this magnitude than Locke, who originated this "underlaborer" metaphor, could be neutral between hylomorphism and corpuscularianism. But, here again, when I say that neutrality is not a desideratum, I am not saying this from a neutral philosophical perspective. I am not laying foundations for liberalism by claiming that recent Davidsonian philosophy of language or Kuhnian philosophy of science has demonstrated that the philosophers of the past were mistaken in asking for neutrality. I am saying that Kuhn, Davidson, Wittgenstein, and Dewey provide us with redescriptions of familiar phenomena which, taken together, buttress Berlin's way of describing alternative political institutions and theories. These philosophers help provide a redescription for political liberalism, but political liberalism also helps provide a redescription of their activity – one which lets us see that there is no natural order of philosophical

8 I do not mean to suggest resurrecting the distinction between the cognitive and the noncognitive, much less assigning allegiance to social institutions to the latter category. With Davidson, I hold that the distinction between true and false (the positivists' mark of "cognitive status") is as applicable to statements like "Yeats was a great poet," and "Democracy is better than tyranny," as to statements like "The earth goes around the sun." My point about the questions of the form "How do you *know* that . . . ?" which I have listed is simply that there is no practicable way to silence doubt on such matters. Those who press such questions are asking for an epistemic position which nobody is ever likely to have about any matter of moral importance.

inquiry. Nothing requires us to first get straight about language, then about belief and knowledge, then about personhood, and finally about society. There is no such thing as "first philosophy" – neither metaphysics nor philosophy of language nor philosophy of science. But, once again and for the last time, that claim about philosophy itself is just one more terminological suggestion made on behalf of the same cause, the cause of providing contemporary liberal culture with a vocabulary which is all its own, cleansing it of the residues of a vocabulary which was suited to the needs of former days.

I can perhaps make this abjuration of philosophical neutrality in the interest of political liberalism more palatable by referring yet again to the Wittgensteinian analogy between vocabularies and tools. I said in the first chapter that the problem with this comparison is that the person who designs a new tool can usually explain what it will be useful for – why she wants it – in advance; by contrast, the creation of a new form of cultural life, a new vocabulary, will have its utility explained only retrospectively. We cannot see Christianity or Newtonianism or the Romantic movement or political liberalism as a tool while we are still in the course of figuring out how to use it. For there are as yet no clearly formulatable ends to which it is a means. But once we figure out how to use the vocabularies of these movements, we can tell a story of progress, showing how the literalization of certain metaphors served the purpose of making possible all the good things that have recently happened. Further, we can now view all these good things as particular instances of some more general good, the overall end which the movement served. This latter process was Hegel's definition of philosophy: "holding your time in thought." l construe this to mean "finding a description of all the things characteristic of your time of which you most approve, with which you unflinchingly identify, a description which will serve as a description of the end toward which the historical developments which led up to your time were means."

Given this meaning of "philosophy," it follows that as Hegel said, "philosophy paints its gray on gray only when a form of life has grown old." Christianity did not know that its purpose was the alleviation of cruelty, Newton did not know that his purpose was modern technology, the Romantic poets did not know that their purpose was to contribute to the development of an ethical consciousness suitable for the culture of political liberalism. But *we* now know these things, for we latecomers can tell the kind of story of progress which those who are actually making progress cannot. We can view these people as toolmakers rather than discoverers because we have a clear sense of the product which the use of those tools produced. The product is *us* – our conscience, our culture, our form of life. Those who made us possible could not have envisaged what they were making possible, and so could not have described the ends to which their work was a means. But *we* can.

Let me now apply this point to the particular case of the relation between political liberalism and Enlightenment rationalism. This relation was the topic of Horkheimer and Adorno's *Dialectic of Enlightenment.* They pointed out, correctly, I think, that the forces unleashed by the Enlightenment have undermined the Enlightenment's own convictions. What they called the "dissolvant rationality" of Enlightenment has, in the course of the triumph of Enlightenment ideas during the last two centuries, undercut the ideas of "rationality" and of "human nature" which the eighteenth century took for granted. They drew the conclusion that liberalism was now intellectually bankrupt, bereft of philosophical foundations, and that liberal society was morally bankrupt, bereft of social glue.

This inference was a mistake. Horkheimer and Adorno assumed that the terms in which those who begin a historical development described their enterprise remain the terms which describe it correctly, and then inferred that the dissolution of that terminology deprives the results of that development of the right to, or the possibility of, continued existence. This is almost never the case. On the contrary, the terms used by the founders of a new form of cultural life will consist largely in borrowings from the vocabulary of the culture which they are hoping to replace. Only when the new form has grown old, has itself become the target of attacks from the avant-garde, will the terminology of that culture begin to take form. The terminology in which a mature culture compares other cultures invidiously with itself, in which it couches its apologetics, are not likely to be the terms which were used to bring about its birth.

Horkheimer and Adorno give an admirable account of the way in which philosophical foundations of society, which they view as linguistic instruments of domination by the rulers, are undercut by Enlightenment skepticism. As they say:

Ultimately the Enlightenment consumed not just the symbols [of social union] but their successors, universal concepts, and spared no remnant of metaphysics. . . . The situation of concepts in the face of the Enlightenment is like that of men of private means in regard to industrial trusts: none can feel safe.[9]

Among the distinctions which have been unable to withstand this dissolution are "absolute validity vs. relative validity" and "morality as op-

9 Max Horkheimer and Theodor W. Adorno, *Dialectic of Enlightenment,* trans. John Cumming (New York: Seabury Press, 1972), p. 23.

posed to prudence." As Horkheimer and Adorno put it, the spirit of the Enlightenment dictates that "every specific theoretic view succumbs to the destructive criticism that it is only a belief – until even the very notions of spirit, of truth and, indeed, enlightenment itself have become animistic magic."[10] This point can be put in my jargon by saying that every specific theoretic view comes to be seen as one more vocabulary, one more description, one more way of speaking.

Horkheimer and Adorno thought it likely that civilization could not survive this process, and had nothing helpful to suggest except what Ricoeur has aptly dubbed "the hermeneutics of suspicion" – constant awareness that any new theoretical proposal was likely to be one more excuse for maintaining the status quo. They said that "if consideration of the destructive aspect of progress is left to its enemies, blindly prag-matized thought loses its transcending quality and its relation to truth."[11] But they had no suggestions for its friends. They had no utopian vision of a culture which was able to incorporate and make use of an understand-ing of the dissolvant character of rationality, of the self-destructive char-acter of the Enlightenment. They did not try to show how "pragmatized thought" might cease to be blind and become clear-sighted.

Yet various other writers – people who wanted to retain Enlighten-ment liberalism while dropping Enlightenment rationalism – have done just that. John Dewey, Michael Oakeshott, and John Rawls have all helped undermine the idea of a transhistorical "absolutely valid" set of concepts which would serve as "philosophical foundations" of liberalism, but each has thought of this undermining as a way of strengthening liberal institutions. They have argued that liberal institutions would be all the better if freed from the need to defend themselves in terms of such foundations – all the better for not having to answer the question "In what does the privileged status of freedom consist?" All three would happily grant that a circular justification of our practices, a justification which makes one feature of our culture look good by citing still another, or comparing our culture invidiously with others by reference to our own standards, is the only sort of justification we are going to get. I am suggesting that we see such writers as these as the self-canceling and self-fulfilling triumph of the Enlightenment. Their pragmatism is antithetical to Enlightenment rationalism, although it was itself made possible (in good dialectical fashion) only by that rationalism. It can serve as the vocabulary of a mature (de-scientized, de-philosophized) Enlightenment liberalism.

10 Ibid., p. 11.
11 Ibid., p. xiii.

Let me cite a passage apiece from these three authors as reminders of their positions. Dewey echoes Hegel's definition of philosophy when he says:

When it is acknowledged that under disguise of dealing with ultimate reality, philosophy has been occupied with the precious values embedded in social traditions, that it has sprung from a clash of social ends and from a conflict of inherited institutions with incompatible contemporary tendencies, it will be seen that the task of future philosophy is to clarify men's ideas as to the social and moral strifes of their own day.[12]

In his Dewey Lectures, Rawls echoes both Berlin and Dewey when he says:

What justifies a conception of justice is not its being true to an order antecedent and given to us, but its congruence with our deeper understanding of ourselves and our aspirations, and our realization that, given our history and the traditions embedded in our public life, it is the most reasonable doctrine for us.[13]

Finally, Oakeshott writes, in a sentence which Dewey might equally well have written:

A morality is neither a system of general principles nor a code of rules, but a vernacular language. General principles and even rules may be elicited from it, but (like other languages) it is not the creation of grammarians; it is made by speakers. What has to be learned in a moral education is not a theorem such as that good conduct is acting fairly or being charitable, nor is it a rule such as "always tell the truth," but how to speak the language intelligently. . . . It is not a device for formulating judgments about conduct or for solving so-called moral problems, but a practice in terms of which to think, to choose, to act, and to utter.[14]

This quotation from Oakeshott gives me a springboard for explaining why I think that the distinction between morality and prudence, and the term "moral" itself, are no longer very useful. My argument turns on the familiar anti-Kantian claim, which Oakeshott is here taking for granted, that "moral principles" (the categorical imperative, the utilitarian princi-

12 John Dewey, *Reconstruction in Philosophy* (Boston: Beacon Press, 1948), p. 26.
13 John Rawls, "Kantian Constructivism in Moral Theory," *Journal of Philosophy* 77 (1980): 519.
14 Michael Oakeshott, *Of Human Conduct* (Oxford: Oxford University Press, 1975), pp. 78–79.

ple, etc.) only have a point insofar as they incorporate tacit reference to a whole range of institutions, practices, and vocabularies of moral and political deliberation. They are reminders of, abbreviations for, such practices, not justifications for such practices. At best, they are pedagogical aids to the acquistion of such practices. This point is common to Hegel and to such recent critics of conventional moral and legal philosophy as Annette Baier, Stanley Fish, Jeffrey Stout, Charles Taylor, and Bernard Williams.[15] If one accepts this point, one will naturally raise the question "Since the classic Kantian opposition between morality and prudence was formulated precisely in terms of the opposition between an appeal to principle and an appeal to expediency, is there any point in keeping the term 'morality' once we drop the notion of 'moral principle'"?

Oakeshott, following Hegel, suggests the answer: We can keep the notion of "morality" just insofar as we can cease to think of morality as the voice of the divine part of ourselves and instead think of it as the voice of ourselves as members of a community, speakers of a common language. We can keep the morality-prudence distinction if we think of it not as the difference between an appeal to the unconditioned and an appeal to the conditioned but as the difference between an appeal to the interests of our community and the appeal to our own, possibly conflicting, private interests. The importance of this shift is that it makes it impossible to ask the question "Is ours a moral society?" It makes it impossible to think that there is something which stands to my community as my community stands to me, some larger community called "humanity" which has an intrinsic nature. This shift is appropriate for what Oakeshott calls a *societas* as opposed to a *universitas,* to a society conceived as a band of eccentrics collaborating for purposes of mutual protection rather than as a band of fellow spirits united by a common goal.

Oakeshott's answer coincides with Wilfrid Sellars's thesis that morality is a matter of what he calls "we-intentions," that the core meaning of "immoral action" is "the sort of thing *we* don't do."[16] An immoral action is, on this account, the sort of thing which, if done at all, is done only by animals, or by people of other families, tribes, cultures, or historical epochs. If done by one of us, or if done repeatedly by one of us, that person ceases to be one of us. She becomes an outcast, someone who

---

15 It is also, of course, a point familiar from Marx and the Marxists. Unfortunately, however, the point is distorted in these authors by a fuzzy distinction between "ideology" and a form of thought (the Marxists' own) which escapes being "ideology." On the uselessness of the notion of "ideology," see Raymond Geuss, *The Idea of a Critical Theory* (Cambridge: Cambridge University Press, 1981).

16 See Wilfrid Sellars, *Science and Metaphysics* (London: Routledge & Kegan Paul, 1968), chaps. 6 and 7. I return to this point in Chapter 9.

doesn't speak our language, even though she may once have appeared to do so. On Sellars's account, as on Hegel's, moral philosophy takes the form of an answer to the question "Who are 'we', how did we come to be what we are, and what might we become?" rather than an answer to the question "What rules should dictate my actions?" In other words, moral philosophy takes the form of historical narration and utopian speculation rather than of a search for general principles.

This Oakeshott-Sellars way of looking at morality as a set of practices, *our* practices, makes vivid the difference between the conception of morality as the voice of a divinized portion of our soul, and as the voice of a contingent human artifact, a community which has grown up subject to the vicissitudes of time and chance, one more of Nature's "experiments." It makes clear why the morality-prudence distinction breaks down when we attempt to transfer it to questions about whether the glue that holds our society together is "moral" or "prudential" in nature. The distinction only makes sense for individuals. It would make sense for societies only if "humanity" had a nature over and above the various forms of life which history has thrown up so far. But if the demands of a morality are the demands of a language, and if languages are historical contingencies, rather than attempts to capture the true shape of the world or the self, then to "stand unflinchingly for one's moral convictions" is a matter of identifying oneself with such a contingency.

Let me now try to put this point together with my earlier claim that the heroes of liberal society are the strong poet and the utopian revolutionary. Such a synthesis will seem paradoxical, and doomed, if one thinks of the poet or the revolutionary as "alienated." But the paradox begins to vanish if one drops an assumption which lurks behind many recent uses of the term "alienation." This is the idea that those who are alienated are people who are protesting in the name of humanity against arbitrary and inhuman social restrictions. One can substitute for this the idea that the poet and the revolutionary are protesting in the name of the society itself against those aspects of the society which are unfaithful to its own self-image.

This substitution seems to cancel out the difference between the revolutionary and the reformer. But one can define the *ideally* liberal society as one in which this difference *is* canceled out. A liberal society is one whose ideals can be fulfilled by persuasion rather than force, by reform rather than revolution, by the free and open encounters of present linguistic and other practices with suggestions for new practices. But this is to say that an ideal liberal society is one which has no purpose except freedom, no goal except a willingness to see how such encounters go and to abide by the outcome. It has no purpose except to make life easier for poets and revolutionaries while seeing to it that they make life harder for others only by words, and not deeds. It is a society whose hero is the strong poet and the revolutionary because it recognizes that it is what it is, has the morality it has, speaks the language it does, not because it approximates the will of God or the nature of man but because certain poets and revolutionaries of the past spoke as they did.

To see one's language, one's conscience, one's morality, and one's highest hopes as contingent products, as literalizations of what once were accidentally produced metaphors, is to adopt a self-identity which suits one for citizenship in such an ideally liberal state. That is why the ideal citizen of such an ideal state would be someone who thinks of the founders and the preservers of her society as such poets, rather than as people who had discovered or who clearly envisioned the truth about the world or about humanity. She herself may or may not be a poet, may or may not find her own metaphors for her own idiosyncratic fantasies, may or may not make those fantasies conscious. But she will be commonsensically Freudian enough to see the founders and the transformers of society, the acknowledged legislators of her language and thus of her morality, as people who did happen to find words to fit their fantasies, metaphors which happened to answer to the vaguely felt needs of the rest of the society. She will be commonsensically Bloomian enough to take for granted that it is the revolutionary artist and the revolutionary scientist, not the academic artist or the normal scientist, who most clearly exemplifies the virtues which she hopes her society will itself embody.

To sum up, the citizens of my liberal utopia would be people who had a sense of the contingency of their language of moral deliberation, and thus of their consciences, and thus of their community. They would be liberal ironists – people who met Schumpeter's criterion of civilization, people who combined commitment with a sense of the contingency of their own commitment. I shall conclude this chapter by trying to bring this figure of the liberal ironist into sharper focus by contrasting my view with that of two philosophers with whom I have wide areas of agreement, but whose views differ from mine in opposite ways. To put the differences crudely: Michel Foucault is an ironist who is unwilling to be a liberal, whereas Jürgen Habermas is a liberal who is unwilling to be an ironist.

Both Foucault and Habermas are, like Berlin, critics of the traditional Platonic and Kantian attempts to isolate a core component of the self. Both see Nietzsche as critically important. Foucault thinks of Nietzsche as having taught him to avoid the attempt at a suprahistorical perspective, the attempt to find timeless origins – to be satisfied with a genealogical narrative of contingencies.[17] Nietzsche also taught him to look twice at liberalism – to look behind the new freedoms which political democracy has brought, at new forms of constraint which democratic societies have imposed.

But whereas Foucault takes Nietzsche as an inspiration, Habermas, though agreeing with Nietzschean criticisms of the "subject-centered reason" of traditional rationalism, sees him as leading us to a dead end. Habermas thinks of Nietzsche as making clear the bankruptcy, for purposes of human "emancipation," of what Habermas calls the "philosophy of subjectivity" (roughly, the attempt to spin moral obligation out of our own vitals, to find deep within us, beyond historical contingencies and the accidents of socialization, the origins of our responsibility to others). With Nietzsche, Habermas says, "The criticism of modernity [i.e., the attempt to come to terms with loss of the kinds of social cohesion found in premodern societies][18] dispenses for the first time with its retention of an emancipatory content."[19] Habermas takes this refusal of the attempt to emancipate to be Nietzsche's legacy to Heidegger, Adorno, Derrida, and Foucault – a disastrous legacy, one which has made philosophical reflection at best irrelevant, and at worst antagonistic, to liberal hope. Habermas thinks of these thinkers – theorists devoured by their own irony – as constituting a sort of reductio ad absurdum of the philosophy of subjectivity.

Habermas's own response to Nietzsche is to try to undercut Nietzsche's attack on our religious and metaphysical traditions by replacing the "philosophy of subjectivity" with a "philosophy of intersubjectivity" – replacing the old "subject-centered conception of 'reason'" shared by Kant and Nietzsche with what Habermas calls "communicative reason." Habermas here makes the same move as Sellars: Both philosophers try to construe reason as the internalization of social norms, rather than as a built-in component of the human self. Habermas wants to "ground" democratic institutions in the same way as Kant hoped to – but to do the job better, by invoking a notion of "domination-free communication" to replace "re-

17 See Foucault's "Nietzche, Genealogy History" in his *Language, Counter-Memory, Practice: Selected Essays and Interviews,* ed. Donald F. Bouchard (Ithaca, N.Y.: Cornell University Press, 1977), esp. pp. 146, 152–153.

18 See Jürgen Habermas, *The Philosophical Discourse of Modernity,* trans. Frederick Lawrence (Cambridge, Mass.: MIT Press, 1987), p. 139: "Since the close of the eighteenth century, the discourse of modernity has had a single theme under ever new titles: the weakening of the forces of social bonding, privatization and diremption – in short the deformations of a one-sidedly rationalized everyday praxis which evoke the need for something equivalent to the unifying power of religion."

19 Ibid., p. 94.

spect for human dignity" as the aegis under which society is to become more cosmopolitan and democratic.

Foucault's response to attempts such as those of Habermas, Dewey, and Berlin – attempts to build a philosophy around the needs of a democratic society – is to point out the drawbacks of this society, the ways in which it does *not* allow room for self-creation, for private projects. Like Habermas and Sellars, he accepts Mead's view that the self is a creation of society. Unlike them, he is not prepared to admit that the selves shaped by modern liberal societies are better than the selves earlier societies created. A large part of Foucault's work – the most valuable part, in my view – consists in showing how the patterns of acculturation characteristic of liberal societies have imposed on their members kinds of constraints of which older, premodern societies had not dreamed. He is not, however, willing to see these constraints as compensated for by a decrease in pain, any more than Nietzsche was willing to see the resentfulness of "slave-morality" as compensated for by such a decrease.

My disagreement with Foucault amounts to the claim that this decrease does, in fact, compensate for those constraints. I agree with Habermas that Foucault's account of how power has shaped our contemporary subjectivity "filters out all the aspects under which the eroticization and internalization of subjective nature also meant a gain in freedom and expression."[20] More important, I think that contemporary liberal society already contains the institutions for its own improvement – an improvement which can mitigate the dangers Foucault sees. Indeed, my hunch is that Western social and political thought may have had the last *conceptual* revolution it needs.[21] J. S. Mill's suggestion that governments devote themselves to optimizing the balance between leaving people's private lives alone and preventing suffering seems to me pretty much the last word. Discoveries about who is being made to suffer can be left to the workings of a free press, free universities, and enlightened public opinion – enlightened, for example, by books like *Madness and Civiliza-*

---

20 Habermas, *The Philosophical Discourse of Modernity*, p. 292. Habermas's complaint echoes those of Michael Walzer and Charles Taylor. See their essays in *Foucault: A Critical Reader*, ed. David Couzens Hoy (Oxford: Blackwell, 1986). I make similar complaints in "Moral Identity and Private Autonomy," in *Foucault*, ed. François Ewald (Paris: Editions du Seuil) (in press).

21 This is, of course, not to say that the world has had the last *political* revolution it needs. It is hard to imagine a diminution of cruelty in countries like South Africa, Paraguay, and Albania without violent revolution. But in such countries, raw courage (like that of the leaders of COSATU or the signers of Charta 77) is the relevant virtue, not the sort of reflective acumen which makes contributions to social theory. In such places the sort of "unmasking" which Foucault is so good at is irrelevant. For there power swaggers naked, and nobody is under any illusions.

tion and *Discipline and Punish*, as well as those like *Germinal, Black Boy, The Road to Wigan Pier*, and *1984*.

Foucault, however, shares with Marx and Nietzsche the conviction that we are too far gone for reform to work – that a convulsion is needed, that our imagination and will are so limited by the socialization we have received that we are unable even to propose an alternative to the society we have now.[22] He was not willing to think of himself as speaking as a member of *any* "we," much less use "we liberals" as I have been doing. As he said,

I do not appeal to any "we" – to any of those "we"'s whose consensus, whose values, whose traditions constitute the framework for a thought and define the conditions in which it can be validated. But the problem is, precisely, to decide if it is actually suitable to place oneself within a "we" in order to assert the principles one recognizes and the values one accepts; or if it is not, rather, necessary to make the future formation of a "we" possible, by elaborating the question.[23]

This is, indeed, the problem. But I disagree with Foucault about whether in fact it is necessary to form a new "we." My principal disagreement with him is precisely over whether "we liberals" is or is not good enough.[24]

Foucault would not appreciate my suggestion that his books can be assimilated into a liberal, reformist political culture. I think that part of the explanation for his reaction would be that despite his agreement with Mead, Sellars, and Habermas that the self, the human subject, is simply whatever acculturation makes of it, he still thinks in terms of something deep within human beings, which is deformed by acculturation. One bit of evidence for this claim is that Foucault is notably reluctant to grant that (as I shall be arguing in Chapter 4) there is no such thing as the "language of the oppressed." Occasionally he suggests that he is speaking

22 Foucault once said in an interview, "I think to imagine another system is to extend our participation in the present system" *(Language, Counter-Memory, Practice,* p. 230).
23 *The Foucault Reader,* ed. Paul Rabinow (New York: Pantheon, 1984), p. 385. The quotation is from a conversation with Rabinow.
24 I agree with Foucault that the constitution of a new "we" can, indeed, result from asking the right question. A community of intellectuals was constituted, in the seventeenth century, by Galileo's question "Is any motion more 'natural' than any other?" Another was constituted by Marx's question "Is the state more than the executive committee of the bourgeoisie?" But forming new communities is no more an end in itself than is political revolution. Expanding the range of our present "we," on the other hand, is one of the two projects which an ironist liberal takes to be ends in themselves, the other being self-invention. (But by "end in itself," of course, she means only "project which I cannot imagine defending on the basis of noncircular argument.")

"for" the insane, or that his work reveals "subjugated knowledges . . . blocks of historical knowledge which were present but disguised within the body of functionalist and systematizing theory."[25] Many passages in Foucault, including the one about "we" quoted above, exemplify what Bernard Yack has called the "longing for total revolution," and the "demand that our autonomy be embodied in our institutions."[26] It is precisely this sort of yearning which I think should, among citizens of a liberal democracy, be reserved for private life. The sort of autonomy which self-creating ironists like Nietzsche, Derrida, or Foucault seek is not the sort of thing that *could* ever be embodied in social institutions. Autonomy is not something which all human beings have within them and which society can release by ceasing to repress them. It is something which certain particular human beings hope to attain by self-creation, and which a few actually do. The desire to be autonomous is not relevant to the liberal's desire to avoid cruelty and pain – a desire which Foucault shared, even though he was unwilling to express it in those terms.

Most ironists confine this longing to the private sphere, as (or so I argue in Chapter 5) Proust did and as Nietzsche and Heidegger should have done. Foucault was not content with this sphere. Habermas ignores it, as irrelevant to his purposes. The compromise advocated in this book amounts to saying: *Privatize* the Nietzschean-Sartrean-Foucauldian attempt at authenticity and purity, in order to prevent yourself from slipping into a political attitude which will lead you to think that there is some social goal more important than avoiding cruelty.

So much, then, for my disagreements with Foucault's attempt to be an ironist without being a liberal. My disagreements with Habermas's attempt to be a liberal without being an ironist become obvious when one realizes how deeply Habermas would dislike my claim that a liberal utopia would be a *poeticized* culture. Habermas sees my aestheticizing

---

25 Michel Foucault, *Power/Knowledge: Selected Interviews and Other Writings 1972–77*, ed. Colin Gordon (Brighton: Harvester Press, 1980), p. 82. Habermas comments on this passage (*The Philosophical Discourse of Modernity*, pp. 279–280). I agree with him that it exemplifies Foucault's attempt to avoid self-referential difficulties by "singling out his genealogy from all the rest of the human sciences in a manner that is reconcilable with the fundamental assumptions of his own theory." I also agree that the attempt fails.

26 Bernard Yack, *The Longing for Total Revolution: Philosophic Sources of Social Discontent from Rousseau to Marx and Nietzsche* (Princeton, N.J.: Princeton University Press, 1986), p. 385. Yack makes a very good case for his claim that the idea of something deeply human which society has deformed comes from Rousseau by way of Kant's attempt to see a portion of the self as outside of nature. Sellars's naturalization of the obligation-benevolence distinction, like Mead's view of the self, helps one pull up the roots of the temptation – typical of contemporary radicalism – to see "society" as *intrinsically* dehumanizing.

talk of metaphor, conceptual novelty, and self-invention as an unfortunate preoccupation with what he calls the "world-disclosing function of language" as opposed to its "problem-solving function" within "intramundane praxis." He distrusts the exaltation of the former function which he finds in neo-Nietzschean figures such as Heidegger and Foucault. He thinks Castoriadis's attempt to invoke this function in his *Imaginary Institution of Society* equally dubious.[27]

Habermas is willing to grant the Kuhnian point that "the specific languages of science and technology, law and morality, economics, political science, etc. . . . live off the revivifying power of metaphorical tropes."[28] But he thinks that I go much too far – dangerously far – when I suggest that "science and morality, economics and politics are delivered up to a process of language-creating protuberances *in just the same way* as art and philosophy."[29] He wants world-disclosure always to be checked for "validity" against intramundane practice. He wants there to be *argumentative* practices, conducted within "expert cultures," which cannot be overturned by exciting, romantic disclosures of new worlds. He is more afraid of the sort of "romantic" overthrow of established institutions exemplified by Hitler and Mao than of the suffocating effect of what Dewey called "the crust of convention" (e.g., the possibly suffocating effect of traditional divisions between "spheres of culture"). He is more afraid of those who, like Foucault, wish to see their own autonomy reflected in institutions than he is of what Foucault feared – the ability of the "expert cultures" to exert "biopower."[30]

Habermas's response to both sets of fears is, however, the same. He thinks that the dangers from both sides can be avoided if decisions about changes in public institutions and policies are made through a process of "domination-free communication." This seems to me a good way to restate the traditional liberal claim that the only way to avoid perpetuat-

27 See, for example, Habermas's treatment of Castoriadis's idea of "the self-transparency of a society that does not hide its imaginary origin beneath extrasocietal projections and knows itself explicitly as a self-instituting society" (*The Philosophical Discourse of Modernity*, p. 318). Habermas criticizes both me and Castoriadis for indulging in *Lebensphilosophie;* this charge means, roughly, that we both want to poeticize rather than rationalize. For my own (obviously more sympathetic) view of Castoriadis, see my "Unger, Castoriadis and the Romance of a National Future," *Northwestern University Law Review* (in press).

28 Habermas, *The Philosophical Discourse of Modernity*, p. 209.

29 Ibid., p. 206. The quotation is Habermas's account of the gist of an article of mine called "Deconstruction and Circumvention," *Critical Inquiry* 11 (1984): 1–23 – an account which says that I have let the "sober insights of pragmatism" be beclouded by the "Nietzschean pathos of a *Lebensphilosophie* that has made the linguistic turn."

30 Habermas is, however, by no means oblivious to the latter sort of danger, which he has diagnosed as the "colonization of the life-world." (See his *Theory of Communicative Action*, vol. 2, pp. 391–396.)

ing cruelty within social institutions is by maximizing the quality of education, freedom of the press, educational opportunity, opportunities to exert political influence, and the like. So the difference between Habermas's attempt to reconstruct a form of rationalism and my recommendation that culture should be poeticized is not reflected in any political disagreement. We do not disagree about the worth of traditional democratic institutions, or about the sorts of improvements these institutions need, or about what counts as "freedom from domination." Our differences concern *only* the self-image which a democratic society should have, the rhetoric which it should use to express its hopes. Unlike my political differences with Foucault, my differences with Habermas are what are often called "merely philosophical" differences.

Habermas thinks it essential to a democratic society that its self-image embody the universalism, and some form of the rationalism, of the Enlightenment. He thinks of his account of "communicative reason" as a way of updating rationalism. I want not to update either universalism or rationalism but to dissolve both and replace them with something else. So I see Habermas's substitution of "communicative reason" for "subject-centered reason" as just a misleading way of making the same point I have been urging: A liberal society is one which is content to call "true" (or "right" or "just") whatever the outcome of undistorted communication happens to be, whatever view wins in a free and open encounter. This substitution amounts to dropping the image of a preestablished harmony between the human subject and the object of knowledge, and thus to dropping the traditional epistemological-metaphysical problematic.

Habermas is willing to drop most of that problematic. But even after he has done so, he still insists on seeing the process of undistorted communication as convergent, and seeing that convergence as a guarantee of the "rationality" of such communication. The residual difference I have with Habermas is that his universalism makes him substitute such convergence for ahistorical grounding, whereas my insistence on the contingency of language makes me suspicious of the very idea of the "universal validity" which such convergence is supposed to underwrite. Habermas wants to preserve the traditional story (common to Hegel and to Peirce) of asymptotic approach to *foci imaginarii*. I want to replace this with a story of increasing willingness to live with plurality and to stop asking for universal validity. I want to see freely arrived at agreement as agreement on how to accomplish common purposes (e.g., prediction and control of the behavior of atoms or people, equalizing life-chances, decreasing cruelty), but I want to see these common purposes against the background of an increasing sense of the radical diversity of private purposes, of the radically poetic character of individual lives, and of the

67

merely poetic foundations of the "we-consciousness" which lies behind our social institutions.

Abandoning universalism is my way of doing justice to the claims of the ironists whom Habermas distrusts: Nietzsche, Heidegger, Derrida. Habermas looks at these men from the point of view of public needs. I agree with Habermas that as *public* philosophers they are at best useless and at worst dangerous, but I want to insist on the role they and others like them can play in accommodating the ironist's *private* sense of identity to her liberal hopes. All that is in question, however, is accommodation – not synthesis. My "poeticized" culture is one which has given up the attempt to unite one's private ways of dealing with one's finitude and one's sense of obligation to other human beings.

For Habermas, however, this compartmentalization of the self, this division of one's final vocabulary into two independent parts, is itself objectionable. To him, such a compartmentalization looks like a concession to irrationalism, an attempt to grant rights to an "other to reason." In my view, however, the purported opposition between reason and its other (e.g., the passions, Nietzsche's will to power, Heidegger's Being) is one we can abandon when we abandon the notion that "reason" names a healing, reconciling, unifying power – the source of human solidarity. If there is no such source, if the idea of human solidarity is simply the fortunate happenstance creation of modern times, then we no longer need a notion of "communicative reason" to substitute for that of "subject-centered reason." We do not need to replace religion with a philosophical account of a healing and unifying power which will do the work once done by God.

I should like to replace both religious and philosophical accounts of a suprahistorical ground or an end-of-history convergence with a historical narrative about the rise of liberal institutions and customs – the institutions and customs which were designed to diminish cruelty, make possible government by the consent of the governed, and permit as much domination-free communication as possible to take place. Such a narrative would clarify the conditions in which the idea of truth as correspondence to reality might gradually be replaced by the idea of truth as what comes to be believed in the course of free and open encounters. That shift from epistemology to politics, from an explanation of the relation between "reason" and reality to an explanation of how political freedom has changed our sense of what human inquiry is good for, is a shift which Dewey was willing to make but from which Habermas hangs back. Habermas still wants to insist that "the transcendent moment of *universal* validity bursts every provinciality asunder . . . the validity laid claim to is distinguished from the social currency of a de facto established practice and yet serves it as the foundation of an existing consensus." It is precisely this claim of universal validity which what I have called the "contingency of language" makes implausible, and which the poeticized culture of my liberal utopia would no longer make. Such a culture would instead agree with Dewey that "imagination is the chief instrument of the good . . . art is more moral than moralities. For the latter either are, or tend to become, consecrations of the status quo. . . . The moral prophets of humanity have always been poets even though they spoke in free verse or by parable."[31]

31 John Dewey, *Art as Experience* (New York: Capricorn Books, 1958), p. 348.

PART II

Ironism and theory

4

Private irony and liberal hope

All human beings carry about a set of words which they employ to justify their actions, their beliefs, and their lives. These are the words in which we formulate praise of our friends and contempt for our enemies, our long-term projects, our deepest self-doubts and our highest hopes. They are the words in which we tell, sometimes prospectively and sometimes retrospectively, the story of our lives. I shall call these words a person's "final vocabulary."

It is "final" in the sense that if doubt is cast on the worth of these words, their user has no noncircular argumentative recourse. Those words are as far as he can go with language; beyond them there is only helpless passivity or a resort to force. A small part of a final vocabulary is made up of thin, flexible, and ubiquitous terms such as "true," "good," "right," and "beautiful." The larger part contains thicker, more rigid, and more parochial terms, for example, "Christ," "England," "professional standards," "decency," "kindness," "the Revolution," "the Church," "progressive," "rigorous," "creative." The more parochial terms do most of the work.

I shall define an "ironist" as someone who fulfills three conditions: (1) She has radical and continuing doubts about the final vocabulary she currently uses, because she has been impressed by other vocabularies, vocabularies taken as final by people or books she has encountered; (2) she realizes that argument phrased in her present vocabulary can neither underwrite nor dissolve these doubts; (3) insofar as she philosophizes about her situation, she does not think that her vocabulary is closer to reality than others, that it is in touch with a power not herself. Ironists who are inclined to philosophize see the choice between vocabularies as made neither within a neutral and universal metavocabulary nor by an attempt to fight one's way past appearances to the real, but simply by playing the new off against the old.

I call people of this sort "ironists" because their realization that anything can be made to look good or bad by being redescribed, and their renunciation of the attempt to formulate criteria of choice between final vocabularies, puts them in the position which Sartre called "meta-stable": never quite able to take themselves seriously because always aware

73

that the terms in which they describe themselves are subject to change, always aware of the contingency and fragility of their final vocabularies, and thus of their selves. Such people take naturally to the line of thought developed in the first two chapters of this book. If they are also liberals – people for whom (to use Judith Shklar's definition) "cruelty is the worst thing they do" – they will take naturally to the views offered in the third chapter.

The opposite of irony is common sense. For that is the watchword of those who unselfconsciously describe everything important in terms of the final vocabulary to which they and those around them are habituated. To be commonsensical is to take for granted that statements formulated in that final vocabulary suffice to describe and judge the beliefs, actions and lives of those who employ alternative final vocabularies. People who pride themselves on common sense will find the line of thought developed in Part I distasteful.

When common sense is challenged, its adherents respond at first by generalizing and making explicit the rules of the language game they are accustomed to play (as some of the Greek Sophists did, and as Aristotle did in his ethical writings). But if no platitude formulated in the old vocabulary suffices to meet an argumentative challenge, the need to reply produces a willingness to go beyond platitudes. At that point, conversation may go Socratic. The question "What is *x*?" is now asked in such a way that it cannot be answered simply by producing paradigm cases of *x*-hood. So one may demand a definition, an essence.

To make such Socratic demands is not yet, of course, to become an ironist in the sense in which I am using this term. It is only to become a "metaphysician," in a sense of that term which I am adapting from Heidegger. In this sense, the metaphysician is someone who takes the question "What is the intrinsic nature of (e.g., justice, science, knowledge, Being, faith, morality, philosophy)?" at face value. He assumes that the presence of a term in his own final vocabulary ensures that it refers to something which *has* a real essence. The metaphysician is still attached to common sense, in that he does not question the platitudes which encapsulate the use of a given final vocabulary, and in particular the platitude which says there is a single permanent reality to be found behind the many temporary appearances. He does not redescribe but, rather, analyzes old descriptions with the help of other old descriptions.

The ironist, by contrast, is a nominalist and a historicist. She thinks nothing has an intrinsic nature, a real essence. So she thinks that the occurrence of a term like "just" or "scientific" or "rational" in the final vocabulary of the day is no reason to think that Socratic inquiry into the essence of justice or science or rationality will take one much beyond the language games of one's time. The ironist spends her time worrying about the possibility that she has been initiated into the wrong tribe, taught to play the wrong language game. She worries that the process of socialization which turned her into a human being by giving her a language may have given her the wrong language, and so turned her into the wrong kind of human being. But she cannot give a criterion of wrongness. So, the more she is driven to articulate her situation in philosophical terms, the more she reminds herself of her rootlessness by constantly using terms like "Weltanschauung," "perspective," "dialectic," "conceptual framework," "historical epoch," "language game," "redescription," "vocabulary," and "irony."

The metaphysician responds to that sort of talk by calling it "relativistic" and insisting that what matters is not what language is being used but what is *true*. Metaphysicians think that human beings by nature desire to know. They think this because the vocabulary they have inherited, their common sense, provides them with a picture of knowledge as a relation between human beings and "reality," and the idea that we have a need and a duty to enter into this relation. It also tells us that "reality," if properly asked, will help us determine what our final vocabulary should be. So metaphysicians believe that there are, out there in the world, real essences which it is our duty to discover and which are disposed to assist in their own discovery. They do not believe that anything can be made to look good or bad by being redescribed – or, if they do, they deplore this fact and cling to the idea that reality will help us resist such seductions.

By contrast, ironists do not see the search for a final vocabulary as (even in part) a way of getting something distinct from this vocabulary right. They do not take the point of discursive thought to be *knowing*, in any sense that can be explicated by notions like "reality," "real essence," "objective point of view," and "the correspondence of language of reality." They do not think its point is to find a vocabulary which accurately represents something, a transparent medium. For the ironists, "final vocabulary" does not mean "the one which puts all doubts to rest" or "the one which satisfies our criteria of ultimacy, or adequacy, or optimality." They do not think of reflection as being governed by criteria. Criteria, on their view, are never more than the platitudes which contextually define the terms of a final vocabulary currently in use. Ironists agree with Davidson about our inability to step outside our language in order to compare it with something else, and with Heidegger about the contingency and historicity of that language.

This difference leads to a difference in their attitude toward books. Metaphysicians see libraries as divided according to disciplines, corresponding to different objects of knowledge. Ironists see them as divided

75

according to traditions, each member of which partially adopts and partially modifies the vocabulary of the writers whom he has read. Ironists take the writings of all the people with poetic gifts, all the original minds who had a talent for redescription – Pythagoras, Plato, Milton, Newton, Goethe, Kant, Kierkegaard, Baudelaire, Darwin, Freud – as grist to be put through the same dialectical mill. The metaphysicians, by contrast, want to start by getting straight about which of these people were poets, which philosophers, and which scientists. They think it essential to get the genres right – to order texts by reference to a previously determined grid, a grid which, whatever else it does, will at least make a clear distinction between knowledge claims and other claims upon our attention. The ironist, by contrast, would.like to avoid cooking the books she reads by using *any* such grid (although, with ironic resignation, she realizes that she can hardly help doing so).

For a metaphysician, "philosophy," as defined by reference to the canonical Plato–Kant sequence, is an attempt to know about certain things – quite general and important things. For the ironist, "philosophy," so defined, is the attempt to apply and develop a particular antecedently chosen final vocabulary – one which revolves around the appearance-reality distinction. The issue between them is, once again, about the contingency of our language – about whether what the common sense of our own culture shares with Plato and Kant is a tip-off to the way the world is, or whether it is just the characteristic mark of the discourse of people inhabiting a certain chunk of space-time. The metaphysician assumes that our tradition can raise no problems which it cannot solve – that the vocabulary which the ironist fears may be merely "Greek" or "Western" or "bourgeois" is an instrument which will enable us to get at something universal. The metaphysician agrees with the Platonic Theory of Recollection, in the form in which this theory was restated by Kierkegaard, namely, that we have the truth within us, that we have built-in criteria which enable us to recognize the right final vocabulary when we hear it. The cash value of this theory is that our contemporary final vocabularies are close enough to the right one to let us converge upon it – to formulate premises from which the right conclusions will be reached. The metaphysician thinks that although we may not have all the answers, we have already got criteria for the right answers. So he thinks "right" does not merely mean "suitable for those who speak as we do" but has a stronger sense – the sense of "grasping real essence."

For the ironist, searches for a final vocabulary are not destined to converge. For her, sentences like "All men by nature desire to know" or "Truth is independent of the human mind" are simply platitudes used to inculcate the local final vocabulary, the common sense of the West. She is an ironist just insofar as her own final vocabulary does not contain such notions. Her description of what she is doing when she looks for a better final vocabulary than the one she is currently using is dominated by metaphors of making rather than finding, of diversification and novelty rather than convergence to the antecedently present. She thinks of final vocabularies as poetic achievements rather than as fruits of diligent inquiry according to antecedently formulated criteria.

Because metaphysicians believe that we already possess a lot of the "right" final vocabulary and merely need to think through its implications, they think of philosophical inquiry as a matter of spotting the relations between the various platitudes which provide contextual definitions of the terms of this vocabulary. So they think of refining or clarifying the use of terms as a matter of weaving these platitudes (or, as they would prefer to say, these intuitions) into a perspicuous system. This has two consequences. First, they tend to concentrate on the thinner, more flexible, more ubiquitous items in this vocabulary – words like "true," "good," "person," and "object." For the thinner the term, the more platitudes will employ it. Second, they take the paradigm of philosophical inquiry to be logical argument – that is spotting the inferential relationships between propositions rather than comparing and contrasting vocabularies.

The typical strategy of the metaphysician is to spot an apparent contradiction between two platitudes, two intuitively plausible propositions, and then propose a distinction which will resolve the contradiction. Metaphysicians then go on to embed this distinction within a network of associated distinctions – a philosophical theory – which will take some of the strain off the initial distinction. This sort of theory construction is the same method used by judges to decide hard cases, and by theologians to interpret hard texts. That activity is the metaphysician's paradigm of rationality. He sees philosophical theories as converging – a series of discoveries about the nature of such things as truth and personhood, which get closer and closer to the way they really are, and carry the culture as a whole closer to an accurate representation of reality.

The ironist, however, views the sequence of such theories – such interlocked patterns of novel distinctions – as gradual, tacit substitutions of a new vocabulary for an old one. She calls "platitudes" what the metaphysician calls "intuitions." She is inclined to say that when we surrender an old platitude (e.g., "The number of biological species is fixed" or "Human beings differ from animals because they have sparks of the divine with them" or "Blacks have no rights which whites are bound to respect"), we have made a change rather than discovered a fact. The ironist, observing the sequence of "great philosophers" and the interaction between their thought and its social setting, sees a series of changes in the linguistic and other practices of the Europeans. Whereas the metaphysician sees the modern Europeans as particularly good at discovering how things really are, the ironist sees them as particularly rapid in changing their self-image, in re-creating themselves.

The metaphysician thinks that there is an overriding intellectual duty to present arguments for one's controversial views – arguments which will start from relatively uncontroversial premises. The ironist thinks that such arguments – logical arguments – are all very well in their way, and useful as expository devices, but in the end not much more than ways of getting people to change their practices without admitting they have done so. The ironist's preferred form of argument is dialectical in the sense that she takes the unit of persuasion to be a vocabulary rather than a proposition. Her method is redescription rather than inference. Ironists specialize in redescribing ranges of objects or events in partially neologistic jargon, in the hope of inciting people to adopt and extend that jargon. An ironist hopes that by the time she has finished using old words in new senses, not to mention introducing brand-new words, people will no longer ask questions phrased in the old words. So the ironist thinks of logic as ancillary to dialectic, whereas the metaphysician thinks of dialectic as a species of rhetoric, which in turn is a shoddy substitute for logic.

I have defined "dialectic" as the attempt to play off vocabularies against one another, rather than merely to infer propositions from one another, and thus as the partial substitution of redescription for inference. I used Hegel's word because I think of Hegel's *Phenomenology* both as the beginning of the end of the Plato–Kant tradition and as a paradigm of the ironist's ability to exploit the possibilities of massive redescription. In this view, Hegel's so-called dialectical method is not an argumentative procedure or a way of unifying subject and object, but simply a literary skill – skill at producing surprising gestalt switches by making smooth, rapid transitions from one terminology to another.

Instead of keeping the old platitudes and making distinctions to help them cohere, Hegel constantly changed the vocabulary in which the old platitudes had been stated; instead of constructing philosophical theories and arguing for them, he avoided argument by constantly shifting vocabularies, thereby changing the subject. In practice, though not in theory, he dropped the idea of getting at the truth in favor of the idea of making things new. His criticism of his predecessors was not that their propositions were false but that their languages were obsolete. By inventing this sort of criticism, the younger Hegel broke away from the Plato–Kant sequence and began a tradition of ironist philosophy which is continued in Nietzsche, Heidegger, and Derrida. These are the philosophers who define their achievement by their relation to their predecessors rather than by their relation to the truth.

A more up-to-date word for what I have been calling "dialectic" would be "literary criticism." In Hegel's time it was still possible to think of plays, poems, and novels as making vivid something already known, of literature as ancillary to cognition, beauty to truth. The older Hegel thought of "philosophy" as a discipline which, because cognitive in a way that art was not, took precedence over art. Indeed, he thought that this discipline, now that it had attained maturity in the form of his own Absolute Idealism, could and would make art as obsolete as it made religion. But, ironically and dialectically enough, what Hegel actually did, by founding an ironist tradition within philosophy, was help de-cognitivize, de-metaphysize philosophy. He helped turn it into a literary genre.[1] The young Hegel's practice undermined the possibility of the sort of convergence to truth about which the older Hegel theorized. The great commentators on the older Hegel are writers like Heine and Kierkegaard, people who treated Hegel the way we now treat Blake, Freud, D. H. Lawrence, or Orwell.

We ironists treat these people not as anonymous channels for truth but as abbreviations for a certain final vocabulary and for the sorts of beliefs and desires typical of its users. The older Hegel became a name for such a vocabulary, and Kierkegaard and Nietzsche have become names for others. If we are told that the actual lives such men lived had little to do with the books and the terminology which attracted our attention to them, we brush this aside. We treat the names of such people as the names of the heroes of their own books. We do not bother to distinguish Swift from *saeva indignatio*, Hegel from Geist, Nietzsche from Zarathustra, Marcel Proust from Marcel the narrator, or Trilling from The Liberal Imagination. We do not care whether these writers managed to live up to their own self-images.[2] What we want to know is whether to adopt those

---

1 From this point of view, both analytic philosophy and phenomenology were throwbacks to a pre-Hegelian, more or less Kantian, way of thinking – attempts to preserve what I am calling "metaphysics" by making it the study of the "conditions of possibility" of a medium (consciousness, language).

2 See Alexander Nehamas, *Nietzsche: Life as Literature*, p. 234, where Nehamas says that he is not concerned with "the miserable little man who wrote [Nietzsche's books]." Rather he is concerned (p. 8) with Nietzsche's "effort to create an artwork of himself, a literary character who is also a philosopher [which is also] his effort to offer a positive view without falling back into the dogmatic tradition." In the view I am suggesting, Nietzsche may have been the first philosopher to do consciously what Hegel had done unconsciously.

images – to re-create ourselves, in whole or in part, in these people's image. We go about answering this question by experimenting with the vocabularies which these people concocted. We redescribe ourselves, our situation, our past, in those terms and compare the results with alternative redescriptions which use the vocabularies of alternative figures. We ironists hope, by this continual redescription, to make the best selves for ourselves that we can.

Such comparison, such playing off of figures against each other, is the principal activity now covered by the term "literary criticism." Influential critics, the sort of critics who propose new canons – people like Arnold, Pater, Leavis, Eliot, Edmund Wilson, Lionel Trilling, Frank Kermode, Harold Bloom – are not in the business of explaining the real meaning of books, nor of evaluating something called their "literary merit." Rather they spend their time placing books in the context of other books, figures in the context of other figures. This placing is done in the same way as we place a new friend or enemy in the context of old friends and enemies. In the course of doing so, we revise our opinions of both the old and the new. Simultaneously, we revise our own moral identity by revising our own final vocabulary. Literary criticism does for ironists what the search for universal moral principles is supposed to do for metaphysicians.

For us ironists, nothing can serve as a criticism of a final vocabulary save another such vocabulary; there is no answer to a redescription save a re-re-redescription. Since there is nothing beyond vocabularies which serves as a criterion of choice between them, criticism is a matter of looking on this picture and on that, not of comparing both pictures with the original. Nothing can serve as a criticism of a person save another person, or of a culture save an alternative culture – for persons and cultures are, for us, incarnated vocabularies. So our doubts about our own characters or our own culture can be resolved or assuaged only by enlarging our acquaintance. The easiest way of doing that is to read books, and so ironists spend more of their time placing books than in placing real live people. Ironists are afraid that they will get stuck in the vocabulary in which they were brought up if they only know the people in their own neighborhood, so they try to get acquainted with strange people (Alcibiades, Julien Sorel), strange families (the Karamazovs, the Casaubons), and strange communities (the Teutonic Knights, the Nuer, the mandarins of the Sung).

Ironists read literary critics, and take them as moral advisers, simply because such critics have an exceptionally large range of acquaintance. They are moral advisers not because they have special access to moral truth but because they have been around. They have read more books and are thus in a better position not to get trapped in the vocabulary of any single book. In particular, ironists hope that critics will help them perform the sort of dialectical feat which Hegel was so good at. That is, they hope critics will help them continue to admire books which are prima facie antithetical by performing some sort of synthesis. We would like to be able to admire both Blake and Arnold, both Nietzsche and Mill, both Marx and Baudelaire, both Trotsky and Eliot, both Nabokov and Orwell. So we hope some critic will show how these men's books can be put together to form a beautiful mosaic. We hope that critics can resdescribe these people in ways which will enlarge the canon, and will give us a set of classical texts as rich and diverse as possible. This task of enlarging the canon takes the place, for the ironist, of the attempt by moral philosophers to bring commonly accepted moral intuitions about particular cases into equilibrium with commonly accepted general moral principles.[3]

It is a familiar fact that the term "literary criticism" has been stretched further and further in the course of our century. It originally meant comparison and evaluation of plays, poems, and novels – with perhaps an occasional glance at the visual arts. Then it got extended to cover past criticism (for example, Dryden's, Shelley's, Arnold's, and Eliot's prose, as well as their verse). Then, quite quickly, it got extended to the books which had supplied past critics with their critical vocabulary and were supplying present critics with theirs. This meant extending it to theology, philosophy, social theory, reformist political programs, and revolutionary manifestos. In short, it meant extending it to every book likely to provide candidates for a person's final vocabulary.

Once the range of literary criticism is stretched that far there is, of course, less and less point in calling it *literary* criticism. But for accidental historical reasons, having to do with the way in which intellectuals got jobs in universities by pretending to pursue academic specialties, the name has stuck. So instead of changing the term "literary criticism" to something like "culture criticism," we have instead stretched the word "literature" to cover whatever the literary critics criticize. A literary critic in what T. J. Clarke has called the "Trotskyite-Eliotic" culture of New

---

3   I am here borrowing Rawls's notion of "reflective equilibrium." One might say that literary criticism tries to produce such equilibrium between the proper names of writers rather than between propositions. One of the easiest ways to express the difference between "analytic" and "Continental" philosophy is to say that the former sort trades in propositions and the latter in proper names. When Continental philosophy made its appearance in Anglo-American literature departments, in the guise of "literary theory," this was not the discovery of a new method or approach but simply the addition of further names (the names of philosophers) to the range of those among which equilibrium was sought.

York in the '30s and '40s was expected to have read *The Revolution Betrayed* and *The Interpretation of Dreams*, as well as *The Wasteland, Man's Hope,* and *An American Tragedy.* In the present Orwellian-Bloomian culture she is expected to have read *The Gulag Archipelago, Philosophical Investigations,* and *The Order of Things* as well as *Lolita* and *The Book of Laughter and Forgetting.* The word "literature" now covers just about every sort of book which might conceivably have moral relevance – might conceivably alter one's sense of what is possible and important. The application of this term has nothing to do with the presence of "literary qualities" in a book. Rather than detecting and expounding such qualities, the critic is now expected to facilitate moral reflection by suggesting revisions in the canon of moral exemplars and advisers, and suggesting ways in which the tensions within this canon may be eased – or, where necessary, sharpened.

The rise of literary criticism to preeminence within the high culture of the democracies – its gradual and only semiconscious assumption of the cultural role once claimed (successively) by religion, science, and philosophy – has paralleled the rise in the proportion of ironists to metaphysicians among the intellectuals. This has widened the gap between the intellectuals and the public. For metaphysics is woven into the public rhetoric of modern liberal societies. So is the distinction between the moral and the "merely" aesthetic – a distinction which is often used to relegate "literature" to a subordinate position within culture and to suggest that novels and poems are irrelevant to moral reflection. Roughly speaking, the rhetoric of these societies takes for granted most of the oppositions which I claimed (at the beginning of Chapter 3) have become impediments to the culture of liberalism.

This situation has led to accusations of "irresponsibility" against ironist intellectuals. Some of these accusations come from know-nothings – people who have not read the books against which they warn others, and are just instinctively defending their own traditional roles. The know-nothings include religious fundamentalists, scientists who are offended at the suggestion that being "scientific" is not the highest intellectual virtue, and philosophers for whom it is an article of faith that rationality requires the deployment of general moral principles of the sort put forward by Mill and Kant. But the same accusations are made by writers who know what they are talking about, and whose views are entitled to respect. As I have already suggested, the most important of these writers is Habermas, who has mounted a sustained, detailed, carefully argued polemic against critics of the Enlightenment (e.g., Adorno, Foucault) who seem to turn their back on the social hopes of liberal societies. In

Habermas's view, Hegel (and Marx) took the wrong tack in sticking to a philosophy of "subjectivity" – of self-reflection – rather than attempting to develop a philosophy of intersubjective communication.

As I said in Chapter 3, I want to defend ironism, and the habit of taking literary criticism as the presiding intellectual discipline, against polemics such as Habermas's. My defense turns on making a firm distinction between the private and the public. Whereas Habermas sees the line of ironist thinking which runs from Hegel through Foucault and Derrida as destructive of social hope, I see this line of thought as largely irrelevant to public life and to political questions. Ironist theorists like Hegel, Nietzsche, Derrida, and Foucault seem to me invaluable in our attempt to form a private self-image, but pretty much useless when it comes to politics. Habermas assumes that the task of philosophy is to supply some social glue which will replace religious belief, and to see Enlightenment talk of "universality" and "rationality" as the best candidate for this glue. So he sees this kind of criticism of the Enlightenment, and of the idea of rationality, as dissolving the bonds between members of liberal societies. He thinks of the contextualism and perspectivalism for which I praised Nietzsche, in previous chapters, as irresponsible subjectivism.

Habermas shares with the Marxists, and with many of those whom he criticizes, the assumption that the real meaning of a philosophical view consists in its political implications, and that the ultimate frame of reference within which to judge a philosophical, as opposed to a merely "literary," writer, is a political one. For the tradition within which Habermas is working, it is as obvious that political philosophy is central to philosophy as, for the analytic tradition, that philosophy of language is central. But, as I said in Chapter 3, it would be better to avoid thinking of philosophy as a "discipline" with "core problems" or with a social function. It would also be better to avoid the idea that philosophical reflection has a natural starting point – that one of its subareas is, in some natural order of justification, prior to the others. For, in the ironist view I have been offering, there is no such thing as a "natural'" order of justification for beliefs or desires. Nor is there much occasion to use the distinctions between logic and rhetoric, or between philosophy and literature, or between rational and nonrational methods of changing other people's minds.[4] If there is no center to the self, then there are only

4 Where these webs of belief and desire are pretty much the same for large numbers of people, it does become useful to speak of an "appeal to reason" or to "logic," for this simply means an appeal to a widely shared common ground by reminding people of propositions which form part of this ground. More generally, all the traditional metaphysical distinctions can be given a respectable ironist sense by sociologizing them – treating them as distinctions between contingently existing sets of practices, or strategies employed within such practices, rather than between natural kinds.

different ways of weaving new candidates for belief and desire into ante-cendently existing webs of belief and desire. The only important political distinction in the area is that between the use of force and the use of persuasion.

Habermas, and other metaphysicians who are suspicious of a merely "literary" conception of philosophy, think that liberal political freedoms require some consensus about what is universally human. We ironists who are also liberals think that such freedoms require no consensus on any topic more basic than their own desirability. From our angle, all that matters for liberal politics is the widely shared conviction that, as I said in Chapter 3, we shall call "true" or "good" whatever is the outcome of free discussion – that if we take care of political freedom, truth and goodness will take care of themselves.

"Free discussion" here does not mean "free from ideology," but sim-ply the sort which goes on when the press, the judiciary, the elections, and the universities are free, social mobility is frequent and rapid, liter-acy is universal, higher education is common, and peace and wealth have made possible the leisure necessary to listen to lots of different people and think about what they say. I share with Habermas the Peircelike claim that the only general account to be given of our criteria for truth is one which refers to "undistorted communication,"[5] but I do not think there is much to be said about what counts as "undistorted" except "the sort you get when you have democratic political institutions and the conditions for making these institutions function."[6]

The social glue holding together the ideal liberal society described in the previous chapter consists in little more than a consensus that the point of social organization is to let everybody have a chance at self-creation to the best of his or her abilities, and that that goal requires, besides peace and wealth, the standard "bourgeois freedoms." This con-viction would not be based on a view about universally shared human ends, human rights, the nature of rationality, the Good for Man, nor anything else. It would be a conviction based on nothing more profound than the historical facts which suggest that without the protection of something like the institutions of bourgeois liberal society, people will

5   This is not to say that "true" can be defined as "what will be believed at the end of inquiry." For criticism of this Peircian doctrine, see Michael Williams, "Coherence, Justification and Truth," *Review of Metaphysics* 34 (1980): 243–272, and section 2 of my "Pragmatism, Davidson and Truth," in Ernest Lepore, ed. *Truth and Interpretation: Perspectives on the Philosophy of Donald Davidson*, pp. 333–355.

6   In contrast, Habermas and those who agree with him that *Ideologiekritik* is central to philosophy think that there is quite a lot to say. The question turns on whether one thinks that one can give an interesting sense to the word "ideology" – make it mean more than "bad idea."

be less able to work out their private salvations, create their private self-images, reweave their webs of belief and desire in the light of whatever new people and books they happen to encounter. In such an ideal society, discussion of public affairs will revolve around (1) how to balance the needs for peace, wealth, and freedom when conditions require that one of these goals be sacrificed to one of the others and (2) how to equalize opportunities for self-creation and then leave people alone to use, or neglect, their opportunities.

The suggestion that this is all the social glue liberal societies need is subject to two main objections. The first is that as a practical matter, this glue is just not thick enough – that the (predominantly) metaphysical rhetoric of public life in the democracies is essential to the continuation of free institutions. The second is that it is psychologically impossible to be a liberal ironist – to be someone for whom "cruelty is the worst thing we do," and to have no metaphysical beliefs about what all human beings have in common.

The first objection is a prediction about what would happen if ironism replaced metaphysics in our public rhetoric. The second is a suggestion that the public-private split I am advocating will not work: that no one can divide herself up into a private self-creator and a public liberal, that the same person cannot be, in alternate moments, Nietzsche and J. S. Mill.

I want to dismiss the first of these objections fairly quickly, in order to concentrate on the second. The former amounts to the prediction that the prevalence of ironist notions among the public at large, the general adoption of antimetaphysical, antiessentialist views about the nature of morality and rationality and human beings, would weaken and dissolve liberal societies. It is possible that this prediction is correct, but there is at least one excellent reason for thinking it false. This is the analogy with the decline of religious faith. That decline, and specifically the decline of people's ability to take the idea of postmortem rewards seriously, has not weakened liberal societies, and indeed has strengthened them. Lots of people in the eighteenth and nineteenth centuries predicted the opposite. They thought that hope of heaven was required to supply moral fiber and social glue – that there was little point, for example, in having an atheist swear to tell the truth in a court of law. As it turned out, however, willingness to endure suffering for the sake of future reward was transferable from individual rewards to social ones, from one's hopes for paradise to one's hopes for one's grandchildren.[7]

---

7 Hans Blumenberg takes this transfer as central to the development of modern thought and society, and he makes a good case.

The reason liberalism has been strengthened by this switch is that whereas belief in an immortal soul kept being buffeted by scientific discoveries and by philosophers' attempts to keep pace with natural science, it is not clear that any shift in scientific or philosophical opinion could hurt the sort of social hope which characterizes modern liberal societies – the hope that life will eventually be freer, less cruel, more leisured, richer in goods and experiences, not just for our descendants but for everybody's descendants. If you tell someone whose life is given meaning by this hope that philosophers are waxing ironic over real essence, the objectivity of truth, and the existence of an ahistorical human nature, you are unlikely to arouse much interest, much less do any damage. The idea that liberal societies are bound together by philosophical beliefs seems to me ludicrous. What binds societies together are common vocabularies and common hopes. The vocabularies are, typically, parasitic on the hopes – in the sense that the principal function of the vocabularies is to tell stories about future outcomes which compensate for present sacrifices.

Modern, literate, secular societies depend on the existence of reasonably concrete, optimistic, and plausible *political* scenarios, as opposed to scenarios about redemption beyond the grave. To retain social hope, members of such a society need to be able to tell themselves a story about how things might get better, and to see no insuperable obstacles to this story's coming true. If social hope has become harder lately, this is not because the clerks have been committing treason but because, since the end of World War II, the course of events has made it harder to tell a convincing story of this sort. The cynical and impregnable Soviet Empire, the continuing shortsightedness and greed of the surviving democracies, and the exploding, starving populations of the Southern Hemisphere make the problems our parents faced in the 1930s – Fascism and unemployment – look almost manageable. People who try to update and rewrite the standard social democratic scenario about human equality, the scenario which their grandparents wrote around the turn of the century, are not having much success. The problems which metaphysically inclined social thinkers believe to be caused by our failure to find the right sort of theoretical glue – a philosophy which can command wide assent in an individualistic and pluralistic society – are, I think, caused by a set of historical contingencies. These contingencies are making it easy to see the last few hundred years of European and American history – centuries of increasing public hope and private ironism – as an island in time, surrounded by misery, tyranny, and chaos. As Orwell put it, "The democratic vistas seem to end in barbed wire."

I shall come back to this point about the loss of social hope when I discuss Orwell in Chapter 8. For the moment, I am simply trying to disentangle the public question "Is absence of metaphysics politically dangerous?" from the private question "Is ironism compatible with a sense of human solidarity?" To do so, it may help to distinguish the way nominalism and historicism look at present, in a liberal culture whose public rhetoric – the rhetoric in which the young are socialized – is still metaphysical, from the way they might look in a future whose public rhetoric is borrowed from nominalists and historicists. We tend to assume that nominalism and historicism are the exclusive property of intellectuals, of high culture, and that the masses cannot be so blasé about their own final vocabularies. But remember that once upon a time atheism, too, was the exclusive property of intellectuals.

In the ideal liberal society, the intellectuals would still be ironists, although the nonintellectuals would not. The latter would, however, be commonsensically nominalist and historicist. So they would see themselves as contingent through and through, without feeling any particular doubts about the contingencies they happened to be. They would not be bookish, nor would they look to literary critics as moral advisers. But they would be commonsensical nonmetaphysicians, in the way in which more and more people in the rich democracies have been commonsensical nontheists. They would feel no more need to answer the questions "*Why* are you a liberal? Why do you *care* about the humiliation of strangers?" than the average sixteenth-century Christian felt to answer the question "Why are you a Christian?" or than most people nowadays feel to answer the question "Are you saved?"[8] Such a person would not need a justification for her sense of human solidarity, for she was not raised to play the language game in which one asks and gets justifications for that sort of belief. Her culture is one in which doubts about the public rhetoric of the culture are met not by Socratic requests for definitions and principles, but by Deweyan requests for concrete alternatives and programs. Such a culture could, as far as I can see, be every bit as self-critical and every bit as devoted to human equality as our own familiar, and still metaphysical, liberal culture – if not more so.

But even if I am right in thinking that a liberal culture whose public rhetoric is nominalist and historicist is both possible and desirable, I cannot go on to claim that there could or ought to be a culture whose public rhetoric is *ironist*. I cannot imagine a culture which socialized its youth in such a way as to make them continually dubious about their own process of socialization. Irony seems inherently a private matter. On my

---

8 Nietzsche said, with a sneer, "Democracy is Christianity made natural" (*Will to Power*, no. 215). Take away the sneer, and he was quite right.

definition, an ironist cannot get along without the contrast between the final vocabulary she inherited and the one she is trying to create for herself. Irony is, if not intrinsically resentful, at least reactive. Ironists have to have something to have doubts about, something from which to be alienated.

This brings me to the second of the two objections I listed above, and thus to the idea that there is something about being an ironist which unsuits one for being a liberal, and that a simple split between private and public concerns is not enough to overcome the tension.

One can make this claim plausible by saying that there is at least a prima facie tension between the idea that social organization aims at human equality and the idea that human beings are simply incarnated vocabularies. The idea that we all have an overriding obligation to diminish cruelty, to make human beings equal in respect to their liability to suffering, seems to take for granted that there is something within human beings which deserves respect and protection quite independently of the language they speak. It suggests that a nonlinguistic ability, the ability to feel pain, is what is important, and that differences in vocabulary are much less important.

Metaphysics – in the sense of a search for theories which will get at real essence – tries to make sense of the claim that human beings are something more than centerless webs of beliefs and desires. The reason many people think such a claim essential to liberalism is that if men and women were, indeed, nothing more than sentential attitudes – nothing more than the presence or absence of dispositions toward the use of sentences phrased in some historically conditioned vocabulary – then not only human nature, but human *solidarity*, would begin to seem an eccentric and dubious idea. For solidarity with all possible vocabularies seems impossible. Metaphysicians tell us that unless there is some sort of common ur-vocabulary, we have no "reason" not to be cruel to those whose final vocabularies are very unlike ours. A universalistic ethics seems incompatible with ironism, simply because it is hard to imagine stating such an ethic without some doctrine about the nature of man. Such an appeal to real essence is the antithesis of ironism.

So the fact that greater openness, more room for self-creation, is the standard demand made by ironists on their societies is balanced by the fact that this demand seems to be merely for the freedom to speak a kind of ironic theoretical metalanguage which makes no sense to the man in the street. One can easily imagine an ironist badly wanting more freedom, more open space, for the Baudelaires and the Nabokovs, while not giving a thought to the sort of thing Orwell wanted: for example, getting more fresh air down into the coal mines, or getting the Party off the backs of the proles. This sense that the connection between ironism and liberalism is very loose, and that between metaphysics and liberalism pretty tight, is what makes people distrust ironism in philosophy and aestheticism in literature as "elitist."

This is why writers like Nabokov, who claim to despise "topical trash" and to aim at "aesthetic bliss," look morally dubious and perhaps politically dangerous. Ironist philosophers like Nietzsche and Heidegger often look the same, even if we forget about their use by the Nazis. By contrast, even when we are mindful of the use made of Marxism by gangs of thugs calling themselves "Marxist governments," the use made of Christianity by the Inquisition, and the use Gradgrind made of utilitarianism, we cannot mention Marxism, Christianity, or utilitarianism without respect. For there was a time when each served human liberty. It is not obvious that ironism ever has.

The ironist is the typical modern intellectual, and the only societies which give her the freedom to articulate her alienation are liberal ones. So it is tempting to infer that ironists are naturally antiliberal. Lots of people, from Julien Benda to C. P. Snow, have taken a connection between ironism and antiliberalism to be almost self-evident. Nowadays many people take for granted that a taste for "deconstruction" – one of the ironists' current catchwords – is a good sign of lack of moral responsibility. They assume that the mark of the morally trustworthy intellectual is a kind of straightforward, unselfconscious, transparent prose – precisely the kind of prose no self-creating ironist wants to write.

Although some of these inferences may be fallacious and some of these assumptions ungrounded, nevertheless there is something right about the suspicion which ironism arouses. Ironism, as I have defined it, results from awareness of the power of redescription. But most people do not want to be redescribed. They want to be taken on their own terms – taken seriously just as they are and just as they talk. The ironist tells them that the language they speak is up for grabs by her and her kind. There is something potentially very cruel about that claim. For the best way to cause people long-lasting pain is to humiliate them by making the things that seemed most important to them look futile, obsolete, and powerless.[9] Consider what happens when a child's precious possessions – the little things around which he weaves fantasies that make him a little different from all other children – are redescribed as "trash," and thrown away. Or consider what happens when these possessions are made to

9 See Judith Shklar's discussion of humiliation on p. 37 of her *Ordinary Vices* (Cambridge, Mass.: Harvard University Press, 1984) and Ellen Scarry's discussion of the use of humiliation by torturers in chap. 1 of *The Body in Pain*.

look ridiculous alongside the possessions of another, richer, child. Something like that presumably happens to a primitive culture when it is conquered by a more advanced one. The same sort of thing sometimes happens to nonintellectuals in the presence of intellectuals. All these are milder forms of what happened to Winston Smith when he was arrested: They broke his paperweight and punched Julia in the belly, thus initiating the process of making him describe himself in O'Brien's terms rather than his own. The redescribing ironist, by threatening one's final vocabulary, and thus one's ability to make sense of oneself in one's own terms rather than hers, suggests that one's self and one's world are futile, obsolete, *powerless*. Redescription often humiliates.

But notice that redescription and possible humiliation are no more closely connected with ironism than with metaphysics. The metaphysician also redescribes, even though he does it in the name of reason rather than in the name of the imagination. Redescription is a generic trait of the intellectual, not a specific mark of the ironist. So why do ironists arouse *special* resentment? We get a clue to an answer from the fact that the metaphysician typically backs up his redescription with argument – or, as the ironist redescribes the process, disguises his redescription under the cover of argument. But this in itself does not solve the problem, for argument, like redescription, is neutral between liberalism and antiliberalism. Presumably the relevant difference is that to offer an argument in support of one's redescription amounts to telling the audience that they are being *educated*, rather than simply reprogrammed – that the Truth was already in them and merely needed to be drawn out into the light. Redescription which presents itself as uncovering the interlocutor's true self, or the real nature of a common public world which the speaker and the interlocutor share, suggests that the person being redescribed is being empowered, not having his power diminished. This suggestion is enhanced if it is combined with the suggestion that his previous, false, self-description was imposed upon him by the world, the flesh, the devil, his teachers, or his repressive society. The convert to Christianity or Marxism is made to feel that being redescribed amounts to an uncovering of his true self or his real interests. He comes to believe that his acceptance of that redescription seals an alliance with a power mightier than any of those which have oppressed him in the past.

The metaphysician, in short, thinks that there is a connection between redescription and power, and that the right redescription can make us free. The ironist offers no similar assurance. She has to say that our chances of freedom depend on historical contingencies which are only occasionally influenced by our self-redescriptions. She knows of no power of the same size as the one with which the metaphysician claims acquaintance. When she claims that her redescription is better, she cannot give the term "better" the reassuring weight the metaphysician gives it when he explicates it as "in better correspondence with reality."

So I conclude that what the ironist is being blamed for is not an inclination to humiliate but an inability to empower. There is no reason the ironist cannot be a liberal, but she cannot be a "progressive" and "dynamic" liberal in the sense in which liberal metaphysicians sometimes claim to be. For she cannot offer the same sort of social hope as metaphysicians offer. She cannot claim that adopting her redescription of yourself or your situation makes you better able to conquer the forces which are marshaled against you. On her account, that ability is a matter of weapons and luck, not a matter of having truth on your side, or having detected the "movement of history."

There are, then, two differences between the liberal ironist and the liberal metaphysician. The first concerns their sense of what redescription can do for liberalism; the second, their sense of the connection between public hope and private irony. The first difference is that the ironist thinks that the *only* redescriptions which serve liberal purposes are those which answer the question "What humiliates?" whereas the metaphysician also wants to answer the question "Why should I avoid humiliating?" The liberal metaphysician wants our *wish to be kind* to be bolstered by an argument, one which entails a self-redescription which will highlight a common human essence, an essence which is something more than our shared ability to suffer humiliation. The liberal ironist just wants our *chances of being kind,* of avoiding the humiliation of others, to be expanded by redescription. She thinks that recognition of a common susceptibility to humiliation is the *only* social bond that is needed. Whereas the metaphysician takes the morally relevant feature of the other human beings to be their relation to a larger shared power — rationality, God, truth, or history, for example — the ironist takes the morally relevant definition of a person, a moral subject, to be "something that can be humiliated." Her sense of human solidarity is based on a sense of a common danger, not on a common possession or a shared power.

What, then, of the point I made earlier: that people want to be described in their own terms? As I have already suggested, the liberal ironist meets this point by saying that we need to distinguish between redescription for private and for public purposes. For my private purposes, I may redescribe you and everybody else in terms which have nothing to do with my attitude toward your actual or possible suffering. My private purposes, and the part of my final vocabulary which is not relevant to my public actions, are none of your business. But as I am a liberal, the part of my final vocabulary which is relevant to such actions requires me to become aware of all the various ways in which other human beings whom I might act upon can be humiliated. So the liberal ironist needs as much imaginative acquaintance with alternative final vocabularies as possible, not just for her own edification, but in order to understand the actual and possible humiliation of the people who use these alternative final vocabularies.

The liberal metaphysician, by contrast, wants a final vocabulary with an internal and organic structure, one which is not split down the middle by a public-private distinction, not just a patchwork. He thinks that acknowledging that everybody wants to be taken on their own terms commits us to finding a least common denominator of those terms, a single description which will suffice for both public and private purposes, for self-definition and for one's relations with others. He prays, with Socrates, that the inner and the outer man will be as one – that irony will no longer be necessary. He is prone to believe, with Plato, that the parts of the soul and of the state correspond, and that distinguishing the essential from the accidental in the soul will help us distinguish justice from injustice in the state. Such metaphors express the liberal metaphysician's belief that the metaphysical public rhetoric of liberalism must remain central to the final vocabulary of the individual liberal, because it is the portion which expressed what she shares with the rest of humanity – the portion that makes solidarity possible.[10]

But that distinction between a central, shared, obligatory portion and a peripheral, idiosyncratic, optional portion of one's final vocabulary is just the distinction which the ironist refuses to draw. She thinks that what unites her with the rest of the species is not a common language but *just* susceptibility to pain and in particular to that special sort of pain which the brutes do not share with the humans – humiliation. On her conception, human solidarity is not a matter of sharing a common truth or a common goal but of sharing a common selfish hope, the hope that one's world – the little things around which one has woven into one's final vocabulary – will not be destroyed. For public purposes, it does not matter if everybody's final vocabulary is different, as long as there is enough overlap so that everybody has some words with which to express

10 Habermas, for example, attempts to save something of Enlightenment rationalism through a "discourse theory of truth" which will show that the "moral point of view" is a "universal" and "does not express merely the moral intuitions of the average, male, middle-class member of a modern Western society" (Peter Dews, ed., *Autonomy and Solidarity: Interviews with Jürgen Habermas* [London: Verso, 1986]). For the ironist, the fact that nobody had ever had such intuitions before the rise of modern Western societies is quite irrelevant to the question of whether she should share them.

the desirability of entering into other people's fantasies as well as into one's own. But those overlapping words – words like "kindness" or "decency" or "dignity" – do not form a vocabulary which all human beings can reach by reflection on their natures. Such reflection will not produce anything except a heightened awareness of the possibility of suffering. It will not produce a *reason to care* about suffering. What matters for the liberal ironist is not finding such a reason but making sure that she *notices* suffering when it occurs. Her hope is that she will not be limited by her own final vocabulary when faced with the possibility of humiliating someone with a quite different final vocabulary.

For the liberal ironist, skill at imaginative identification does the work which the liberal metaphysician would like to have done by a specifically moral motivation – rationality, or the love of God, or the love of truth. The ironist does not see her ability to envisage, and desire to prevent, the actual and possible humiliation of others – despite differences of sex, race, tribe, and final vocabulary – as more real or central or "essentially human" than any other part of herself. Indeed, she regards it as an ability and a desire which, like the ability to formulate differential equations, arose rather late in the history of humanity and is still a rather local phenomenon. It is associated primarily with Europe and America in the last three hundred years. It is not associated with any power larger than that embodied in a concrete historical situation, for example, the power of the rich European and American democracies to disseminate their customs to other parts of the world, a power which was enlarged by certain past contingencies and has been diminished by certain more recent contingencies.

Whereas the liberal metaphysician thinks that the good liberal knows certain crucial propositions to be true, the liberal ironist thinks the good liberal has a certain kind of know-how. Whereas he thinks of the high culture of liberalism as centering around theory, she thinks of it as centering around literature (in the older and narrower sense of that term – plays, poems, and, especially, novels). He thinks that the task of the intellectual is to preserve and defend liberalism by backing it up with some true propositions about large subjects, but she thinks that this task is to increase our skill at recognizing and describing the different sorts of little things around which individuals or communities center their fantasies and their lives. The ironist takes the words which are fundamental to metaphysics, and in particular to the public rhetoric of the liberal democracies, as just another text, just another set of little human things. Her ability to understand what it is like to make one's life center around these words is not distinct from her ability to grasp what it is like to make

93

one's life center around the love of Christ or of Big Brother. Her liberalism does not consist in her devotion to those particular words but in her ability to grasp the function of many different sets of words.

These distinctions help explain why ironist philosophy has not done, and will not do, much for freedom and equality. But they also explain why "literature" (in the older and narrower sense), as well as ethnography and journalism, is doing a lot. As I said earlier, pain is nonlinguistic: It is what we human beings have that ties us to the nonlanguage-using beasts. So victims of cruelty, people who are suffering, do not have much in the way of a language. That is why there is no such things as the "voice of the oppressed" or the "language of the victims." The language the victims once used is not working anymore, and they are suffering too much to put new words together. So the job of putting their situation into language is going to have to be done for them by somebody else. The liberal novelist, poet, or journalist is good at that. The liberal theorist usually is not.

The suspicion that ironism in philosophy has not helped liberalism is quite right, but that is not because ironist philosophy is inherently cruel. It is because liberals have come to expect philosophy to do a certain job – namely, answering questions like "Why not be cruel?" and "Why be kind?" – and they feel that any philosophy which refuses this assignment must be heartless. But that expectation is a result of a metaphysical upbringing. If we could get rid of the expectation, liberals would not ask ironist philosophy to do a job which it cannot do, and which it defines itself as unable to do.

The metaphysician's association of theory with social hope and of literature with private perfection is, in an ironist liberal culture, reversed. Within a liberal metaphysical culture the disciplines which were charged with penetrating behind the many private appearances to the one general common reality – theology, science, philosophy – were the ones which were expected to bind human beings together, and thus to help eliminate cruelty. Within an ironist culture, by contrast, it is the disciplines which specialize in thick description of the private and idiosyncratic which are assigned this job. In particular, novels and ethnographies which sensitize one to the pain of those who do not speak our language must do the job which demonstrations of a common human nature were supposed to do. Solidarity has to be constructed out of little pieces, rather than found already waiting, in the form of an ur-language which all of us recognize when we hear it.

Conversely, within our increasingly ironist culture, philosophy has become more important for the pursuit of private perfection rather than for any social task. In the next two chapters, I shall claim that ironist philosophers are private philosophers – philosophers concerned to intensify the irony of the nominalist and the historicist. Their work is ill-suited to public purposes, of no use to liberals qua liberals. In Chapters 7 and 8, I shall offer examples of the way in which novelists can do something which is socially useful – help us attend to the springs of cruelty in ourselves, as well as to the fact of its occurrence in areas where we had not noticed it.

5

Self-creation and affiliation:
Proust, Nietzsche, and Heidegger

To illustrate my claim that, for us ironists, theory has become a means to private perfection rather than to human solidarity, I shall discuss some paradigms of ironist theory: the young Hegel, Nietzsche, Heidegger, and Derrida. I shall use the word "theorist" rather than "philosopher" because the etymology of "theory" gives me the connotations I want, and avoids some I do not want. The people I shall be discussing do not think that there is anything called "wisdom" in any sense of the term which Plato would have recognized. So the term "lover of wisdom" seems inappropriate. But *theoria* suggests taking a view of a large stretch of territory from a considerable distance, and this is just what the people I shall be discussing do. They all specialize in standing back from, and taking a large view of, what Heidegger called the "tradition of Western metaphysics" – what I have been calling the "Plato–Kant canon."

The items in this canon, the works of the great metaphysicians, are the classic attempts to see everything steadily and see it whole. The metaphysicians attempt to rise above the plurality of appearances in the hope that, seen from the heights, an unexpected unity will become evident – a unity which is a sign that something *real* has been glimpsed, something which stands behind the appearances and produces them. By contrast, the ironist canon I want to discuss is a series of attempts to look back on the attempts of the metaphysicians to rise to these heights, and to see the unity which underlies the plurality of these attempts. The ironist theorist distrusts the metaphysician's metaphor of a vertical view downward. He substitutes the historicist metaphor of looking back on the past along a horizontal axis. But what he looks back on is not things in general but a very special sort of person, writing a very special kind of book. The topic of ironist theory is metaphysical theory. For the ironist theorist, the story of belief in, and love of, an ahistorical wisdom is the story of successive attempts to find a final vocabulary which is not just the final vocabulary of the individual philosopher but a vocabulary final in every sense – a vocabulary which is no mere idiosyncratic historical product but the last word, the one to which inquiry and history have converged, the one which renders further inquiry and history superfluous.

The goal of ironist theory is to understand the metaphysical urge, the urge to theorize, so well that one becomes entirely free of it. Ironist theory is thus a ladder which is to be thrown away as soon as one has figured out what it was that drove one's predecessors to theorize.[1] The last thing the ironist theorist wants or needs is a theory of ironism. He is not in the business of supplying himself and his fellow ironists with a method, a platform, or a rationale. He is just doing the same thing which all ironists do – attempting autonomy. He is trying to get out from under inherited contingencies and make his own contingencies, get out from under an old final vocabulary and fashion one which will be all his own. The generic trait of ironists is that they do not hope to have their doubts about their final vocabularies settled by something larger than themselves. This means that their criterion for resolving doubts, their criterion of private perfection, is autonomy rather than affiliation to a power other than themselves. All any ironist can measure success against is the past – not by living up to it, but by redescribing it in his terms, thereby becoming able to say, "Thus I willed it."

The generic task of the ironist is the one Coleridge recommended to the great and original poet: to create the taste by which he will be judged. But the judge the ironist has in mind is himself. He wants to be able to sum up his life in his own terms. The perfect life will be one which closes in the assurance that the last of his final vocabularies, at least, really was wholly *his*. The specific difference which distinguishes the ironist *theorist* is simply that his past consists in a particular, rather narrowly confined, literary tradition – roughly, the Plato–Kant canon, and footnotes to that canon. What he is looking for is a redescription of that canon which will cause it to lose the power it has over him – to break the spell cast by reading the books which make up that canon. (Metaphysically, and so misleadingly, put: The ironist wants to find philosophy's secret, true, magical, name – a name whose use will make philosophy one's servant rather than one's master.) The relation of the ironist theorist to the rest

---

1   The motto of ironist theorizing was provided by the old Heidegger, who ended his 1962 lecture "Time and Being" by saying, "A regard to metaphysics still prevails even in the intention to overcome metaphysics. Therefore our task is to cease all overcoming, and leave metaphysics to itself" (*On Time and Being*, trans. Joan Stambaugh [New York: Harper & Row, 1972], p. 24). Heidegger is vividly aware of a possibility which was eventually actualized in the work of Derrida – that Heidegger would be treated as he himself treated Nietzsche, as one more (the last) rung in a ladder which must be cast away. For an example of this awareness, see his repudiation of the "French" idea that his work is continuous with Hegel's, and his denial that there is such a thing as "Heidegger's philosophy" ("Summary of a Seminar," ibid., p. 48). See also the passage in "A Dialogue on Language Between a Japanese and an Inquirer" about the danger of words which were meant as hints and signs (*Winke und Gebärden*) being construed as *concepts*, instruments for grasping something other than themselves (*Zeichen und Chiffren*) (*On the Way to Language*, trans. Peter Hertz [New York: Harper & Row, 1971], pp. 24–27).

of ironist culture, unlike the relation of the metaphysician to the rest of metaphysical culture, is not that of the abstract to the concrete, the general problem to the special cases. It is simply a matter of which concrete things one is ironic about – which items make up the relevant past. The past, for the ironist, is the books which have suggested that there might be such a thing as an unironizable vocabulary, a vocabulary which could not be replaced by being redescribed. Ironist theorists can be thought of as literary critics who specialize in those books – in that particular literary genre.

In our increasingly ironist culture, two figures are often cited as having achieved the sort of perfection Coleridge described: Proust and Nietzsche. Alexander Nehamas, in his recent book on Nietzsche, has brought these two figures together. He points out that they had in common not only the fact that they spent their lives replacing inherited with self-made contingencies, but described themselves as doing exactly that. Both were aware that that very process of self-creation was itself a matter of contingencies of which they would not be unable to be fully conscious, but neither was troubled by the metaphysician's questions about the relation between freedom and determinism. Proust and Nietzsche are paradigm nonmetaphysicians because they so evidently cared only about how they looked to themselves, not how they looked to the universe. But whereas Proust took metaphysics as just one more form of life, it obsessed Nietzsche. Nietzsche was not only a nonmetaphysician, but an antimetaphysical theorist.

Nehamas cites a passage in which Proust's narrator says that he believes:

. . . that in fashioning a work of art we are by no means free, that we do not choose how we shall make it but that it pre-exists and therefore we are obliged, since it is both necessary and hidden, to do what we should have to do if it were a law of nature, that is to say to discover it.

Nehamas comments:

Yet this discovery, which [Proust] explicitly describes as "the discovery of our true life," can be made only in the very process of creating the work of art which describes and constitutes it. And the ambiguous relation between discovery and creation, which matches exactly Nietzsche's own view, also captures perfectly the tension in the very idea of being able to become who one actually is.[2]

2 Alexander Nehamas, *Nietzsche: Life as Literature*, p. 188.

In the sense Nietzsche gave to the phrase, "who one actually is" does not mean "who one actually was all the time" but "whom one turned oneself into in the course of creating the taste by which one ended up judging oneself." The term "ended up" is, however, misleading. It suggests a predestined resting place. But the process of becoming aware of one's causes by redescribing them is bound to be still going on at one's death. Any last, deathbed self-redescription will itself have had causes which there will be no time to redescribe. It will have been dictated by a law of nature one had no time left to discover (but upon which one's strong, admiring critics may some day stumble).

A metaphysician like Sartre may describe the ironist's pursuit of perfection as a "futile passion," but an ironist like Proust or Nietzsche will think that this phrase begs the crucial question. The topic of futility would arise only if one were trying to surmount time, chance, and self-redescription by discovering something more powerful than any of these. For Proust and Nietzsche, however, there is *nothing* more powerful or important than self-redescription. They are not trying to surmount time and chance, but to use them. They are quite aware that what counts as resolution, perfection, and autonomy will always be a function of when one happens to die or to go mad. But this relativity does not entail futility. For there is no big secret which the ironist hopes to discover, and which he might die or decay before discovering. There are only little mortal things to be rearranged by being redescribed. If he had been alive or sane longer, there would have been more material to be rearranged, and thus different redescriptions, but there would never have been the right description. For although the thoroughgoing ironist can use the notion of a "better description," he has no criterion for the application of this term and so cannot use the notion of "the right description." So he sees no futility in his failure to become an *être-en-soi*. The fact that he never wanted to be one, or at least wanted not to want to be one, is just what separates him from the metaphysician.

Despite these similarities between Proust and Nietzsche, there is a decisive difference, and that difference is crucial for my purposes. Proust's project has little to do with politics; like Nabokov, he uses the public issues of the day only for local color. By contrast, Nietzsche often speaks as though he had a social mission, as if he had views relevant to public action – distinctively antiliberal views. But, as also in the case of Heidegger, this antiliberalism seems adventitious and idiosyncratic – for the kind of self-creation of which Nietzsche and Heidegger are models seems to have nothing in particular to do with questions of social policy. I think that a comparison of both men with Proust may help clarify the situation, and also help buttress the claim I made at the end of Chapter 4, namely, that the ironist's final vocabulary can be and should be split into a large private and a small public sector, sectors which have no particular relation to one another.

As a first crude way of blocking out a difference between Proust and Nietzsche we can note that Proust became who he was by reacting against and redescribing people – real live people whom he had met in the flesh – whereas Nietzsche reacted against and redescribed people he had met in books. Both men wanted to create themselves by writing a narrative about the people who had offered descriptions of them; they wanted to become autonomous by redescribing the sources of heteronomous descriptions. But Nietzsche's narrative – the narrative encapsulated in the section called "How the 'True World' Became a Fable" in *The Twilight of the Idols* – describes not persons but, rather, the vocabularies for which certain famous names serve as abbreviations.

The difference between people and ideas is, however, only superficial. What is important is that whereas the collection of people whom Proust met, who described him and whom he redescribed in his novel – parents, servants, family friends, fellow students, duchesses, editors, lovers – is *just* a collection, just the people whom Proust happened to bump into. The vocabularies Nietzsche discusses, by contrast, are linked dialectically, related internally to one another. They are not a chance collection but a dialectical progression, one which serves to describe the life of somebody who is not Friedrich Nietzsche but somebody much bigger. The name Nietzsche most often gives to this big person is "Europe." In the life of Europe, unlike that of Nietzsche, chance does not intrude. As in the young Hegel's *Phenomenology of Spirit* and again in Heidegger's History of Being, there is no room for contingency in the narrative.

Europe, Spirit, and Being are not just accumulations of contingencies, products of chance encounters – the sort of thing Proust knew himself to be. This invention of a larger-than-self hero, in terms of whose career they define the point of their own, is what sets Hegel, Nietzsche, and Heidegger apart from Proust and makes them *theorists* rather than novelists: people who are looking at something large, rather than constructing something small. Although they are genuine ironists, not metaphysicians, these three writers are not yet full-fledged nominalists, because they are not content to arrange little things. They also want to describe a big thing.

That is what sets their narratives apart from *Remembrance of Things Past*. Proust's novel is a network of small, interanimating contingencies. The narrator might never have encountered another madeleine. The newly impoverished Prince de Guermantes did not have to marry Madame Verdurin: He might have found some other heiress. Such contingencies make sense only in retrospect – and they make a different sense every time redescription occurs. But in the narratives of ironist theory, Plato *must* give way to Saint Paul, and Christianity to Enlightenment. A Kant *must* be followed by a Hegel, and a Hegel by a Marx. That is why ironist theory is so treacherous, so liable to self-deception. It is one reason why each new theorist accuses his predecessors of having been metaphysicians in disguise.

Ironist theory must be narrative in form because the ironist's nominalism and historicism will not permit him to think of his work as establishing a relation to real essence; he can only establish a relation to the past. But, unlike other forms of ironist writing – and in particular unlike the ironist novel of which Proust's is paradigmatic – this relation to the past is a relation not to the author's idiosyncratic past but to a larger past, the past of the species, the race, the culture. It is a relation not to a miscellaneous collection of contingent actualities but to the realm of possibility, a realm through which the larger-than-life hero runs his course, gradually exhausting possibilities as he goes. By a happy coincidence, the culture reached the end of this gamut of possibilities just about the time the narrator himself was born.

The figures I am using as paradigms of ironist theorizing – the Hegel of the *Phenomenology,* the Nietzsche of *Twilight of the Idols,* and the Heidegger of the "Letter on Humanism" – have in common the idea that something (history, Western man, metaphysics – something large enough to have a destiny) has exhausted its possibilities. So now all things must be made new. They are not interested only in making themselves new. They also want to make this big thing new; their own autonomy will be a spin-off from this larger newness. They want the sublime and ineffable, not just the beautiful and novel – something incommensurable with the past, not simply the past recaptured through rearrangement and redescription. They want not just the effable and relative beauty of rearrangement but the ineffable and absolute sublimity of the Wholly Other; they want Total Revolution.[3] They want a way of seeing their past which is incommensurable with all the ways in which the past has described itself. By contrast, ironist novelists are not interested in incommensurability. They are content with mere difference. Private autonomy can be gained by redescribing one's past in a way which had not occurred to the past. It does not require apocalyptic novelty of the sort which ironist theory demands. The ironist who is not a theorist will not be bothered by the thought that his own redescriptions of the past will be grist for his successors' redescriptions; his attitude toward his successors

---

3 See Bernard Yack, *The Longing for Total Revolution,* especially part 3.

is simply "good luck to them." But the ironist theorist cannot imagine any successors, for he is the prophet of a new age, one in which no terms used in the past will have application.

I said toward the end of Chapter 4 that the ironist liberal was interested not in power but only in perfection. The ironist theorist, however, still wants the kind of power which comes from a close relation to somebody very large; this is one reason why he is rarely a liberal. Nietzsche's superman shares with Hegel's World-Spirit and Heidegger's Being the duality attributed to Christ: very man, but, in his ineffable aspect, very God. The Christian doctrine of the Incarnation was essential to Hegel's own account of his project, and it turns up again when Nietzsche begins to imagine himself as the Antichrist and again when Heidegger, the ex-Jesuit novice, starts describing Being both as infinitely gentle and as Wholly Other.

Proust, too, was interested in power, but not in finding somebody larger than himself to incarnate or to celebrate. All he wanted was to get out from under finite powers by making their finitude evident. He did not want to befriend power nor to be in a position to empower others, but simply to free himself from the descriptions of himself offered by the people he had met. He wanted not to be merely the person these other people thought they knew him to be, not to be frozen in the frame of a photograph shot from another person's perspective. He dreaded being, in Sartre's phrase, turned into a thing by the eye of the other (by, for example, St. Loup's "hard look," Charlus's "enigmatic stare").[4] His method of freeing himself from those people – of becoming autonomous – was to redescribe the people who had described him. He drew sketches of them from lots of different perspectives – and in particular from lots of different positions in time – and thus made clear that none of these people occupied a privileged standpoint. Proust became autonomous by explaining to himself why the others were not authorities, but simply fellow contingencies. He redescribed them as being as much a product of others' attitudes toward them as Proust himself was a product of their attitudes toward him.

At the end of his life and his novel, by showing what time had done to these other people, Proust showed what he had done with the time he had. He had written a book, and thus created a self – the author of that book – which these people could not have predicted or even envisaged. He had become as much of an authority on the people whom he knew as his younger self had feared they might be an authority on him. This feat

4 *Remembrance of Things Past*, trans. Charles Scott-Moncrief (New York: Random House, 1934), vol. 1, pp. 571, 576.

enabled him to relinquish the very idea of authority, and with it the idea that there is a privileged perspective from which he, or anyone else, is to be described. It enabled him to shrug off the whole idea of affiliation with a superior power – the sort of affiliation which Charlus offered the young Marcel on their first meeting, and which metaphysicians have traditionally offered their readers, an affiliation designed to make the epigone feel like an incarnation of Omnipotence.

Proust temporalized and finitized the authority figures he had met by seeing them as creatures of contingent circumstance. Like Nietzsche, he rid himself of the fear that there was an antecedent truth about himself, a real essence which others might have detected. But Proust was able to do so without claiming to know a truth which was hidden from the authority figures of his earlier years. He managed to debunk authority without setting himself up as authority, to debunk the ambitions of the powerful without sharing them. He finitized authority figures not by detecting what they "really" were but by watching them become different than they had been, and by seeing how they looked when redescribed in terms offered by still other authority figures, whom he played off against the first. The result of all this finitization was to make Proust unashamed of his own finitude. He mastered contingency by recognizing it, and thus freed himself from the fear that the contingencies he had encountered were more than just contingencies. He turned other people from his judges into his fellow sufferers, and thus succeeded in creating the taste by which he judged himself.

Nietzsche, like Proust and the young Hegel, delighted in his own skill at redescription, his ability to pass back and forth between antithetical descriptions of the same situation. All three were skillful at appearing to be on both sides of a single question while actually shifting perspective, thereby changing the question in between successive answers. All three relished the changes time brings. Nietzsche loved to show that, as he put it, everything which has ever been put forward as a hypothesis about "man" is "basically no more than a statement about man within a *very limited* time span."[5] More generally, he loved showing that every description of anything is relative to the needs of some historically conditioned situation. He and the young Hegel both employed this technique for finitizing the great dead philosophers – the great redescribers whom an ironist who takes up philosophy must himself redescribe, and thereby surpass, if he is to become their equal rather than remain their epigone.

When employed by theorists rather than novelists, however, this strategy of finitization raises an obvious problem – the problem which

5 Nietzsche, *Human, All Too Human*, 2.

Hegel's commentators sum up with the phrase "the end of history." If one defines oneself in terms of one's originality vis-à-vis a set of predecessors, and prides oneself on one's ability to redescribe them even more thoroughly and radically than they redescribed each other, one will eventually start asking the question "and who is going to redescribe *me?*" Because the theorist wants to *see* rather than to *rearrange,* to rise above rather than to manipulate, he has to worry about the so-called problem of self-reference – the problem of explaining his own unprecedented success at redescription in the terms of his own theory. He wants to make clear that because the realm of possibility is now exhausted, nobody can rise above him in the way in which he has risen above everyone else. There is, so to speak, no dialectical space left through which to rise; this is as far as thinking can go. The question "Why should I think, how can I possibly claim, that redescription ends with me?" can also be thought of as the question "How can I end my book?" The *Phenomenology of Spirit* ends on an ambiguous note: Its last lines can be interpreted either as opening up to an indefinitely long future or as looking back on a story that is completed. But, notoriously, the note on which some of Hegel's later books end is "And so Germany became Top Nation, and History came to an End."[6]

Kierkegaard said that if Hegel had prefaced the *Science of Logic* with "This is all just a thought-experiment," he would have been the greatest thinker who ever lived.[7] Striking that note would have demonstrated Hegel's grasp of his own finitude, as well as of everybody else's. It would have privatized Hegel's attempt at autonomy, and repudiated the temptation to think that he had affiliated himself with something larger. It would be charitable and pleasant, albeit unjustified by the evidence, to believe that Hegel deliberately refrained from speculating on the nation which would succeed Germany, and the philosopher who would succeed Hegel, because he wanted to demonstrate his own awareness of his own finitude through what Kierkegaard called "indirect communication" – by an ironic gesture rather than by putting forward a claim. It would be nice to think that he deliberately left the future blank as an invitation to his successors to do to him what he had done to his predecessors, rather than as an arrogant assumption that nothing more could possibly be done. But, however it may have been with Hegel, the problem of how to

---

6   The *Lectures on the History of Philosophy* end as the *Phenomenology* began, with Hegel's *Aufhebung* of Fichte and Schelling and the claim that Spirit, now knowing itself to be absolute, "has reached its goal."

7   Kierkegaard's *Journal,* cited (without page reference) by Walter Lowrie in his notes to Kierkegaard, *Concluding Unscientific Postscript,* trans. David Swenson and Walter Lowrie (Princeton, N.J.: Princeton University Press, 1968), p. 558.

finitize while exhibiting a knowledge of one's own finitude – of satisfying Kierkegaard's demand on Hegel – is *the* problem of ironist theory. It is the problem of how to overcome authority without claiming authority. That problem is the ironist's counterpart to the metaphysicians' problem of bridging the gap between appearance and reality, time and eternity, language and the nonlinguistic.

For nontheorists like Proust, there is no such problem. The narrator of *The Past Recaptured* would not be perturbed by the question "Who is going to redescribe me?" For his job was done once he had put the events of his own life in his own order, made a pattern out of all the little things – Gilberte among the hawthorns, the color of the windows in the Guermantes's chapel, the sound of the name "Guermantes," the two walks, the shifting spires. He knows this pattern would have been different had he died earlier or later, for there would have been fewer or more little things which would have had to be fitted into it. But that does not matter. Proust has no problem of how to avoid being *aufgehoben*. Beauty, depending as it does on giving shape to a multiplicity, is notoriously transitory, because it is likely to be destroyed when new elements are added to that multiplicity. Beauty requires a frame, and death will provide that frame.

By contrast, sublimity is neither transitory, relational, reactive, nor finite. The ironist theorist, unlike the ironist novelist, is continually tempted to try for sublimity, not just beauty. That is why he is continually tempted to relapse into metaphysics, to try for one big hidden reality rather than for a pattern among appearances – to hint at the existence of somebody larger than himself called "Europe" or "History" or "Being" whom he incarnates. The sublime is not a synthesis of a manifold, and so it cannot be attained by redescription of a series of temporal encounters. To try for the sublime is to try to make a pattern out of the entire realm of *possibility*, not just of some little, contingent, actualities. Since Kant, the metaphysical attempt at sublimity has taken the form of attempts to formulate the "necessary conditions of all possible *x*." When philosophers make this transcendental attempt, they start playing for bigger stakes than the sort of private autonomy and private perfection which Proust achieved.

In theory, Nietzsche is not playing this Kantian game. In practice, just insofar as he claims to see deeper rather than differently, claims to be free rather than merely reactive, he betrays his own perspectivism and his own nominalism. He thinks that his historicism will save him from this betrayal, but it does not. For what he itches for is a historical sublime, a future which has broken all relations with the past, and therefore can be linked to the philosopher's redescriptions of the past only by negation. Whereas Plato and Kant had prudently taken this sublimity outside of time altogether, Nietzsche and Heidegger cannot use this dodge. They have to stay in time, but to view themselves as separated from all the rest of time by a decisive event.

This quest for the historical sublime – for proximity to some event such as the closing of the gap between subject and object or the advent of the superman or the end of metaphysics – leads Hegel, Nietzsche, and Heidegger to fancy themselves in the role of the "last philosopher." The attempt to be in this position is the attempt to write something which will make it impossible for one to be redescribed except in one's own terms – make it impossible to become an element in anybody else's beautiful pattern, one more little thing. To try for the sublime is to try not just to create the taste by which one judges oneself, but to make it impossible for anybody else to judge one by any other taste. Proust would have been quite content to think that he might serve as an element in other people's beautiful patterns. It pleased him to think that he might play for some successor the role which one of his own precursors – say, Balzac or Saint-Simon – had played for him. But that thought is, sometimes, more than an ironist theorist like Nietzsche can bear.

Consider the contrast between Nietzsche's defenses of "perspectivalism" and his polemics against "reactiveness." As long as he is busy relativizing and historicizing his predecessors, Nietzsche is happy to redescribe them as webs of relations to historical events, social conditions, their own predecessors, and so on. At these moments he is faithful to his own conviction that the self is not a substance, and that we should drop the whole idea of "substance" – of something that cannot be perspectivalized because it has a real essence, a privileged perspective on itself. But at other moments, the moments when he is imagining a superman who will not be just a bundle of idiosyncratic reactions to past stimuli, but will be *pure* self-creation, pure spontaneity, he forgets all about his perspectivalism. When he starts explaining how to be wonderful and different and unlike anything that has ever existed, he talks about human selves as if they were reservoirs of something called "will to power." The superman has an immense reservoir of this stuff, and Nietzsche's own is presumably pretty big. Nietzsche the perspectivalist is interested in finding a perspective from which to look back on the perspectives he inherited, in order to see a beautiful pattern. That Nietzsche can be modeled, as Nehamas models him, on Proust; he can be seen as having created himself as the author of his books. But Nietzsche the theorist of the will to power – the Nietzsche whom Heidegger attacked as "the last metaphysician" – is as interested as Heidegger himself was in getting beyond all perspectives. He wants sublimity, not just beauty.

If Nietzsche had been able to think of the canon of great dead philosophers in the way that Proust thought of the people he happened to meet, he would not have been tempted to be a theorist, would not have striven for sublimity, would have escaped Heidegger's criticism and lived up to Kierkegaard's and Nehamas's expectations. He would have been a Kierkegaard without Christianity, one who remained self-consciously "aesthetic," in Kierkegaard's sense of that term. If he had been faithful to his own perspectivalism and antiessentialism, he would have avoided the temptation into which Hegel fell. That was the temptation of thinking that once you have found a way to subsume your predecessors under a general idea you have thereby done something more than found a redescription of them – a redescription which has proved useful for your own purposes of self-creation. If you go on to conclude that you have found a way to make yourself quite different from those predecessors, to do something quite different from what they did, then you are doing what Heidegger called "relapsing into metaphysics." For now you are claiming that none of the descriptions that applied to them applies to you – that you are separated from them by an abyss. You are acting as if a redescription of one's predecessors got one in touch with a power other than oneself – something capitalized: Being, Truth, History, Absolute Knowledge, or the Will to Power. This was Heidegger's point when he called Nietzsche "merely an inverted Platonist": The same urge to affiliate with somebody bigger which had led Plato to reify "Being" led Nietzsche to try to affiliate himself with "Becoming" and "Power."

Proust had no such temptation. At the end of his life, he saw himself as looking back along a temporal axis, watching colors, sounds, things, and people fall into place from the perspective of his own most recent description of them. He did not see himself as looking down upon the sequence of temporal events from above, as having ascended from a perspectival to a nonperspectival mode of description. *Theoria* was no part of his ambition; he was a perspectivalist who did not have to worry about whether perspectivalism was a true theory. So the lesson I draw from Proust's example is that novels are a safer medium than theory for expressing one's recognition of the relativity and contingency of authority figures. For novels are usually about people – things which are, unlike general ideas and final vocabularies, quite evidently time-bound, embedded in a web of contingencies. Since the characters in novels age and die – since they obviously share the finitude of the books in which they occur – we are not tempted to think that by adopting an attitude toward them we have adopted an attitude toward every *possible* sort of person. By contrast, books which are about ideas, even when written by historicists like Hegel and Nietzsche, look like descriptions of eternal relations between eternal objects, rather than genealogical accounts of the filiation of final vocabularies, showing how these vocabularies were engendered by haphazard matings, by who happened to bump into whom.[8]

The contrast I have been drawing between Proust and Nietzsche raises the central problem which Heidegger tried to solve: namely, how can we write a historical narrative about metaphysics – about successive attempts to find a redescription of the past which the future will not be able to redescribe – without ourselves becoming metaphysicians? How can we tell a historical narrative which ends with oneself without looking as ridiculous as Hegel made himself look? How can one be a theorist – write a narrative of ideas rather than people – which does not pretend to a sublimity which one's own narrative rules out?

Although Nietzsche is full of talk about a "new day," a "new way," a "new soul," a "new man," his eagerness to burst the limits set by the past is sometimes mitigated by his rueful awareness of Hegel's pratfall, and more generally by his sense of the disadvantages of too much historical consciousness for life. Nietzsche realizes that somebody who wants to create himself cannot afford to be too Apollonian. In particular, he cannot imitate Kant's attempt to survey the entire realm of possibility from above. For the idea of a fixed, unchangeable "realm of possibility" is hard to combine with the idea that one might, by one's own efforts, enlarge that realm – not simply take one's place within a predetermined scheme, but change the scheme. An ironist theorist is caught in a dilemma between saying he has actualized the last possibility left open and saying that he has created not just a new actuality but new possibilities. The demands of theory require him to say the former, the demands of self-creation require him to say the latter.

Nietzsche is a confusing, instructive study in the tension between these two demands – a tension which comes out in the strain the attempt to see himself as world-historical put on Nietzsche's boundless sense of humor – the strain wrought on his desire to be utterly novel by his realization that that ambition was by now rather old-fashioned. But the humorless Heidegger is nevertheless the figure who tells us most about this tension. He was far more self-conscious and explicit about the dilemma I have just put forward than either Hegel or Nietzsche, and indeed was obsessed with it. It is not much of an exaggeration to say that the resolution of this dilemma gradually became, in the course of the 1930s, Heidegger's central concern.

8   There are, of course, novels like Thomas Mann's *Doktor Faustus* in which the characters are simply dressed-up generalities. The novel form cannot by itself *insure* a perception of contingency. It only makes it a bit harder to avoid this perception.

In the mid-'20s, Heidegger could still quite unselfconsciously project his own problem about how to be an ironist theorist onto something large ("Dasein") by identifying "guilt" (in a deep "ontological" sense) with the fact that one had not created oneself. "Dasein as such is guilty," he tells us. For Dasein is continually pursued by the "call of conscience," which reminds it that it is being "pursued" by its own uncanniness [*Unheimlichkeit*], the uncanniness which is "the basic kind of Being-in-the-world, even though in an everyday way it has been covered up."⁹ Authenticity is the recognition of this uncanniness. It is achieved only by those who realize that they are "thrown" – realize that they cannot (at least not *yet*) say to the past, "Thus I willed it."

For Heidegger – early and late – what one is is the practices one engages in, and especially the language, the final vocabulary, one uses. For that vocabulary determines what one can take as a possible project. So to say that Dasein is guilty is to say that it speaks somebody else's language, and so lives in a world it never made – a world which, just for this reason, is not its *Heim*. It is guilty because its final vocabulary is just something which it was thrown into – the language that happened to be spoken by the people among whom it grew up. Most people would not feel guilty about this, but people with the special gifts and ambitions shared by Hegel, Proust, and Heidegger *do*. So the simplest answer to the question "what does Heidegger mean by the word 'Dasein'?" is "people like himself" – people who are unable to stand the thought that they are not their own creations. These are the people who immediately see the point of Blake's exclamation, "I must Create a System, or be enslav'd by another Man's."¹⁰ Or, more exactly, such people are "*authentic* Dasein" – Dasein that knows it is Dasein, that it is only *contingently* there where it is, speaking as it does.

Heidegger seems seriously to have thought, when he was writing *Being and Time,* that he was carrying out a transcendental project, namely, giving an accurate list of the "ontological" conditions of possibility of merely

9 See *Sein und Zeit,* 15th ed. (Tübingen: Max Nieweyer, 1979), p. 285, for "Das Dasein als solche ist schuldig." See p. 277 for the claim that "Unheimlichkeit ist die obzwar alltäglich verdeckte Grundart des In-der-Welt-seins," and p. 274. for the "Ruf des Gewissens" (which re-appears in later Heidegger as the *Stimme des Seins*). The equivalent pages of MacQuarrie and Robinson's translation (*Being and Time,* New York: Harper & Row, 1962) are, respectively, 331, 322 and 319. For a good discussion of these sections of *Being and Time,* see John Richardson, *Existential Epistemology* (Oxford: Oxford University Press, 1986), pp. 128–135. Richardson says (p. 132), "The impossibility of an utter self-creation, this sense in which we can never be a 'cause of ourselves,' is the first nullity [*Nichtigkeit* – Heidegger's word for the lack which makes us guilty] in what Heidegger calls 'guilt.'"
10 *Jerusalem,* plate 10, line 20. The next line reads, "I will not Reason and Compare; my business is to Create."

"ontic" states. He seems genuinely to have believed that the ordinary states of mind and life plans of nonintellectuals were "grounded" on the ability of people like himself and Blake to have spectacularly different anxieties and projects. (He tells us with a straight face, for example, that "guilt" as defined above is a condition of the possibility of, for instance, feeling guilty because you have not yet paid back a monetary debt.) Just as Kant seems never to have asked himself how, given the restrictions on human cognition the *Critique of Pure Reason* had discerned, it was possible to assume the "transcendental standpoint" from which that book was purportedly written, so the Heidegger of this period never looks into the question of methodological self-reference. He never asks himself how "ontology" of the sort he was busy producing was, given its own conclusions, possible.

In remarking on this early unselfconsciousness, I am not trying to denigrate Heidegger's early (internally inconsistent, hastily written, brilliantly original) book. Heidegger was, after all, not the first philosopher to have taken his own idiosyncratic spiritual situation for the essence of what it was to be a human being. (The first clear case of a philosopher who did so is Plato, the first Western philosopher whose works survive.) Rather, I am pointing out that there were excellent reasons for Heidegger, in the course of the 1930s, to cease using the words Dasein, "ontology," and "phenomenology" and to stop talking about the "conditions of possibility" of various familiar emotions and situations. He had excellent reason to stop talking as if his subject was something like "the kind of thing all human beings really are, deep down" and to start talking explicitly about what really bothered him: his own particular, private indebtedness to particular past philosophers, his own fear that their vocabularies might have enslaved him, his terror that he would never succeed in creating himself.

From the time when he becomes preoccupied with Nietzsche (who had barely gotten a look-in in *Being and Time*) until his death, Heidegger concentrates on the question "How can I avoid being one more metaphysician, one more footnote to Plato?" The first answer he gives to this question is to change his description of what he wants to write from "phenomenological ontology" to "the history of Being" – the history of a few dozen thinkers, people who created themselves, and the ensuing ages of the world, by embodying a new "understanding of Being" (*Seinsverständis*). All these thinkers were metaphysicians in that they all invoked some form of the Greek appearance-reality distinction: They all envisaged themselves as getting closer to something (the Real) which already awaited them. Even Nietzsche, treated (as Heidegger insisted on treating him) as the theorist of the will to power as ultimate reality, was a metaphysician – albeit "the last metaphysician" because the one who performed the only remaining transformation of Plato: inverting him, so as to make the Real consist in that with which Plato had identified Appearance.[11]

This redescription of the past, and in particular of Nietzsche – this description of the West as the place where Platonism inverted itself and ended up as the will to power – enabled Heidegger to picture himself as a thinker of a new kind. He wanted to be neither a metaphysician nor an ironist, but to combine the advantages of both. He spent much of his time giving the pejorative sense to the term "metaphysics" which Derrida picked up from him and popularized – the sense I have been employing in this book. But he also spent a lot of time being scornful of the aestheticist, pragmatist light-mindedness of the ironists. He thought of them as dilettantish chatterers who lacked the high seriousness of the great metaphysicians – their special relation to Being. As a Schwarzwald redneck, he had an ingrained dislike of North German cosmopolitan mandarins. As a philosopher, he viewed the rise of the ironist intellectuals – many of them Jews – as symptomatic of the degeneracy of what

---

11 Habermas takes the switch from "phenomenological ontology" to "history of Being" to be a result of Heidegger's involvement with the Nazis. In *The Philosophical Discourse of Modernity* he says, "I suspect that Heidegger could find his way to the temporalized *Ursprungsphilosophie* of the later period only by way of his temporary identification with the National Socialist Movement – to whose inner truth and greatness he still attested in 1935" (p. 155). Later he says that Heidegger wanted to blame his blindness to the nature of the Nazi movement on "a sublimated history promoted to the lofty heights of ontology. Thus was born the concept of the history of Being" (p. 159). But a great deal of the story that Heidegger told in the late '30s and the '40s about the history of Being was prefigured in his 1927 lectures on "The Basic Problems of Phenomenology," and presumably would have made up part two of *Being and Time* if that book had ever been completed. Even if the Nazis had not come to power, and if Heidegger had never dreamed of becoming Hitler's *eminence grise*, I suspect that the "turn" would still have been taken.

One important feature of the history of Being which is *not* present in the 1920s is the claim that it "exhausts its possibilities" with Nietzsche. So my hunch is that crucial details of the "sublimated history promoted to the lofty heights of ontology" were fixed not on the day Heidegger asked himself, "How will I look to history in my Nazi uniform?" but, rather, on the day he asked himself, "Will history see me as just one more disciple of Nietzsche's?" Habermas is certainly right that Heidegger needed to find some excuse for his Nazism, and that he wove a (thoroughly unconvincing) self-exculpation into the story he proceeded to tell. But, on my view, it is a story which he would have written anyway, even if he had had less to excuse himself for.

On the general question of the relation between Heidegger's thought and his Nazism, I am not persuaded that there is much to be said except that one of the century's most original thinkers happened to be a pretty nasty character. He was the sort of man who could betray his Jewish colleagues for the sake of his own ambition, and then manage to forget what he had done. But if one holds the view of the self as centerless which I put forward in Chapter 2, one will be prepared to find the relation between the intellectual and the moral virtues, and the relation between a writer's books and other parts of his life, contingent.

he called "the age of the world-picture." He thought the ironist culture of our century, the high culture in which Proust and Freud are central figures, merely the unThinking self-satisfaction of a postmetaphysical nihilism. So he wanted to find a way of being neither metaphysical nor aestheticist. He wanted to see metaphysics as the true and fateful destiny of Europe, rather than simply brushing it off (as both Proust and Freud did). But he also wanted to insist that metaphysics, and therefore Europe, were now over, for – now that Plato had been fully inverted – metaphysics had exhausted its possibilities.

For Heidegger, this task presented itself as the task of how to work within a final vocabulary while somehow simultaneously "bracketing" that vocabulary – to keep the seriousness of its finality while letting it itself express its own contingency. He wanted to construct a vocabulary which would both constantly dismantle itself and constantly take itself seriously. Hegel's and Nietzsche's historicist perspectivisms had led in the direction of this problem, but Hegel had ducked it by talking as if Absolute Knowledge were just around the corner, and as if language were just a dispensable "mediation" which the final union of Subject and Object would supersede. Nietzsche, in turn, had fobbed us off with the suggestion that the superman would somehow get along without *any* vocabulary. Nietzsche vaguely suggests that the child who, in Zarathustra's parable, succeeds the lion (who, in turn, had succeeded the camel) will somehow have all the advantages of thought with none of the disadvantages of speaking some particular language.

Heidegger, to his credit, did not duck the problem; he did not abandon nominalism in favor of nonlinguistic ineffability when the chips were down. Instead, he made a daring, outrageous suggestion about what philosophy might be in an ironist age. In *Being and Time* is a sentence which, I think, describes his ambition both before and after the *Kehre:* "The ultimate business of philosophy is to preserve the *force of the most elementary words* in which Dasein expresses itself, and to keep the common understanding from leveling them off to that unintelligibility which functions . . . as a source of pseudo-problems" (*Being and Time*, p. 262).

The first set of candidates for "most elementary words" Heidegger offered were words like *Dasein, Sorge,* and *Befindlichkeit,* in the new uses which he assigned them when writing *Being and Time.* During the *Kehre,* in the course of (what I imagine to have been) his gradual realization that the jargon of that book, and its transcendental pretensions, made it a tempting target for Kierkegaardian and Nietzschean ridicule, he offered a second set of candidates: the emblematic words used by the great dead metaphysicians – words like *noein, physis,* and *substantia,* redefined by Heidegger to suit his purpose of showing that all these metaphysicians had, appearances to the contrary, been trying to express a sense of Da-sein's finitude.[12] Both sets of words have irony built in – they are all supposed to express authentic Dasein's sense of itself as unable to get along without a final vocabulary while aware that no vocabulary can remain final, its sense of its own "meta-stability." "Dasein" was, so to speak, Heidegger's name for the ironist. But, in his later period, this word is replaced by "Europe" or "the West" – the personification of the place where Being played out a destiny which ended in ironism. For the later Heidegger, to talk about ironism is to talk about the penultimate stage in the story of Europe, the stage which immediately precedes Heidegger and of which he made Nietzsche a symbol – the stage in which "the world becomes view" as the intellectuals (and gradually, everyone else) realize that anything can be made to look good or bad, interesting or boring, by being recontextualized, redescribed.[13]

In my reading of him, all of Heidegger's "most elementary words" are words designed to express the predicament of the ironist theorist – the tension which Nietzsche and Hegel felt but shrugged off, and which Heidegger takes with full seriousness. All these words are supposed to

12 The application of what Heidegger, in what may have been a rare moment of good-humored ruefulness, called "the farfetched and one-sided Heideggerian method of exegesis" (*Introduction to Metaphysics*, trans. Ralph Mannheim [New Haven, Conn.: Yale University Press, 1959], p. 176), always results in the realization that the great philosopher (or poet) whose text is being examined was anticipating *Sein und Zeit*. The text always makes the point that *Sein* and *Dasein* are interlocked, that Being is not something removed and infinite, but rather something which "is there only so long as Dasein is" (see *Sein und Zeit*, p. 212: "Allerdings nur solange Dasein *ist*, das heisst die ontische Möglichkeit von Seinsverständnis, 'gibt es' Sein"; compare *Introduction to Metaphysics*, p. 139, where this is said to be the gist of Fragment 8 of Parmenides).

On the one hand Heidegger wants to say that he and Parmenides (fellow members of the club of Thinkers, as opposed to the epigones who misinterpret and banalize the Thinkers' work) were working along the same lines: "At the very beginning of Western philosophy it became evident that the question of being necessarily embraces the foundations of being-there" (*Introduction to Metaphysics*, p. 174). On the other hand he wants to say that Being itself has changed since Parmenides' time, as a result of growing *Seinsvergessenheit*. He has trouble combining the two claims.

13 See "The Age of the World Picture," in *The Question Concerning Technology and Other Essays*, trans. William Lovitt (New York: Harper & Row, 1977), and especially p. 129: "Hence world picture, when understood essentially, does not mean a picture of the world but the world conceived and grasped as picture. What is, in its entirety, is now taken in such a way that it first is in being and only is in being to the extent that it is set up by man, who represents and sets forth." If, as I do, one forgets about Being and thinks that beings are all there are, then this "humanist" outlook – one which Heidegger himself despised – will be what one understands by Heidegger's claim that "language speaks man" and by his exaltation of "the poetic" as what "opens up worlds." The first three chapters of this book, and especially the exaltation of "the poet" in chapter 3, are an attempt to enlarge upon the idea of "the world as picture" which Heidegger offers in this essay. But, like Derrida, I want to stand Heidegger on his head – to cherish what he loathed.

encapsulate the difficulty of being theoretic and ironic at the same time. Heidegger is thus writing about himself, his own predicament, when he claims to be writing about somebody else's − "Europe's." What binds early to late Heidegger is the hope of finding a vocabulary which will keep him authentic − one which will block any attempt to affiliate oneself with a higher power, to achieve a *ktēma eis aiei*, to escape from time into eternity. He wants words which cannot be "leveled off," which cannot be used as if they were part of the "right" final vocabulary. He wants a self-consuming and continually self-renewing final vocabulary − words which will make clear that they are *not* representations of real essence, *not* ways of getting in touch with a higher power, not themselves instruments of power or means to ends, not attempts to evade Dasein's responsibility of self-creation. He wants words which will, so to speak, do his work for him − relieve him of the tension he feels by taking that tension on themselves. So he has to adopt a view of language which is not only anti-Wittgensteinian but anti-Lockean, a view which has been unfamiliar since speculation about an "Adamic" language petered out in the seventeenth century. For Heidegger, philosophical truth depends upon the very choice of *phonemes*, on the very *sounds* of words.[14]

14 This comes out most clearly in Heidegger's insistence on aurality − an insistence made much of by Derrida, who inverts Heidegger by insisting on "the priority of the written," but who attends to the shapes of written words in the same way that Heidegger attended to the sounds of spoken ones. See, for example, Heidegger's essay on "The Nature of Language," which contains such passages as "When the word is called [by Hölderlin] the mouth's flower and its blossom, we hear the sound of language rising like the earth. From whence? From Saying in which it comes to pass that World is made to appear. The sound rings out in the resounding assembly call which, open to the Open, makes World appear in all things" (*On the Way to Language*, p. 101).

When Heidegger says that "German and Greek are the two philosophical languages" and that Greek "is the only language which is what it says," I take him to be saying that philosophy is untranslatable in the sense in which poetry is said to be − that the *sounds* matter. Unless he thinks this, his endless wordplay, with its invocation of archaic German words, does not make sense. (See, for example, the discussion of *war, wahr,* and *wahren* at *Early Greek Thinking*, trans. David Krell and Frank Capuzzi [New York: Harper & Row, 1975], p. 36). For the question of etymology does not interest Heidegger, and he shrugs off accusations of having traded in false etymologies. All he is interested in is *resonances:* The causal histories which produced, or did not produce, these resonances are irrelevant.

Heidegger wants thought to be "poetized" and wants us to realize that "Strictly, it is language that speaks. Man first speaks when, and only when, he responds to language by listening to its appeal" (*Poetry, Language, Thinking,* trans. Albert Hofstadter [New York: Harper & Row, 1971], p. 216). But he never tells us anything very enlightening about the relation between poets and thinkers − about why, for instance, Sophocles and Hölderlin count as the former and Parmenides and himself as the latter. His envy of Hölderlin is almost palpable, but he is not about to compete with him. No sooner has he said that "we are thinking the same thing that Hölderlin is saying poetically" than he feels it necessary to insert a baffling I-take-it-all-back paragraph which says, for example, "Poetry and thinking meet each other in one and the same only when, and

It is easy to take this view as a reductio ad absurdum of Heidegger's project. But one can see its attractions if one sees the difficulty of the problem Heidegger had set himself: the problem of how to surpass, place, and set aside all past theory without oneself theorizing. In his own jargon, this comes across as the problem of talking about Being without talking about what all beings have in common. (Talking in this latter way is, on Heidegger's definition, the essence of metaphysics.) The problem of whether his own purportedly nontheoretic jargon is all that different from other people's confessedly theoretic jargon comes through for Heidegger as the problem of how "to touch upon the nature of language without doing it injury."[15] More specifically, it is the problem of how to keep "hints and gestures" [*Winke und Gebärden*] distinct from the "signs and chiffres" [*Zeichen und Chiffren*] of metaphysics: how, for example, to prevent the phrase "house of Being" (one of Heidegger's descriptions of language) from being taken as a "mere hasty image which helps us in imagining what we will."[16] The only solution to such problems is: do not put Heidegger's words in any context, do not treat them as movable pieces in a game, or as tools, or as relevant to any questions save Heidegger's own. In short, give his words the privilege you extend to a lyric which you love too much to treat as an object of "literary criticism" – a lyric which you recite, but do not (for fear of injuring it) relate to anything else.

Only if the phonemes, the sounds, matter, does such a plea make sense. For if they did not matter, then we would be free to treat Heidegger's words – the fragments of final vocabulary which he developed for himself – as counters in a language game which people other than himself could get in on. We would be able to treat even terms like *Haus des Seins* in the contextualist way made familiar by Saussure and Wittgenstein, as a more or less useful tool for some purpose which is extraneous to the term itself. But if we do that, we shall eventually be driven back on the questions "What is the point of playing the game in question?" and "For what purpose is this final vocabulary useful?" The only available answer to both seems to be the one Nietzsche gave: It increases our power; it helps us get what we antecedently decided we want.[17]

<hr/>

only as long as, they remain distinctly in the distinctness of their natures" (ibid., p. 218).

Heidegger did not want to be thought a "failed poet" any more than he wanted to be thought the professor who translated Nietzsche into academic jargon. But the former description comes to mind quite readily when reading what he wrote after the war, just as does the latter when one reads certain sections of *Sein und Zeit*.

15 *On the Way to Language*, p. 22.
16 Ibid., p. 26.
17 Heidegger thinks that this view of language became inevitable once Plato dis-

Heidegger thinks that if we are to avoid just this identification of truth with power – to avoid the sort of humanism and pragmatism advocated in this book, forms of thought which he took to be the most degraded versions of the nihilism in which metaphysics culminates – we have to say that final vocabularies are not just means to ends but, indeed, houses of Being. But this claim requires him to poetize philosophical language by letting the phonemes matter, not just the uses to which phonemes are put.

Most contemporary objections to Wittgensteinian views of language and pragmatist views of truth come from "realist" philosophers (e.g., Wilfrid Sellars, Bernard Williams) who argue that physical science is privileged over other parts of discourse. For Heidegger, this gets things exactly the wrong way around. On his view, pragmatism, Wittgenstein, and physical science deserve each other. Heidegger thinks it is poetry, rather than physics, which shows the inadequacy of a language-game account of language. Consider one of his examples of unparaphrasability, the occurrence of the word *ist* in Goethe's line "Über allen Gipfeln, ist Ruh."[18] There seems to be something wrong with trying to construe this *ist* as an instrument for accomplishing a purpose. It can, of course, be so construed, and Heidegger gives us no argument why it should not be. But he wants us to consider the question "Given that there *seems* something wrong with so construing it, what would language have to be if there *were* something wrong?" His answer is that there would have to be certain "elementary words" – words which have "force" apart from their use by what he calls "the common understanding." The common understanding is what a language-game theory catches. But *force* is what the idea of the "house of Being" is supposed to help us catch. If no words had force, there would be no need for philosophy as the attempt to preserve that force.

This account of what Heidegger meant by the phrase "the ultimate business of philosophy" raises the obvious question "How does Heidegger know an elementary word when he sees one, a word that has force rather than just use?" If he is as finite, as bound to time and place, as the rest of us, how can he claim to be able to recognize an elementary word when he hears one without turning back into a metaphysician? We can get a clue to his answer from a line in one of the few poems he published (although, apparently, he wrote many): "Being's poem – man – is just

tinguished between "meaning" and "sensory vehicle of meaning." This inevitability is part of the larger inevitability he sees in the progression of Plato to Nietzsche – the fated deliquescence of the appearance-reality distinction into the your-power-versus-my-power distinction.

18 *Introduction to Metaphysics*, p. 90.

begun." He thinks of man – or rather of European man – as the person whose life has been spent in passing from certain final vocabularies to certain others. So, if you want to pick out elementary words, you write a bildungsroman about a character called "Europe," trying to spot the crucial transitions in Europe's life. Think of Heidegger as doing the sort of thing for "Being's poem" which a critic might try to do for "poetry in English." A sufficiently ambitious critic, such as Bloom, constructs a canon not only of poets but also of poems and of lines within poems, trying to identify just which lines of which poems opened up or closed off options for the successor poets. The "most elemental" lines of English poetry are the ones which determine the historical position in which a twentieth-century poet writing in English finds himself: They are the house in which he lives, not the tools he uses. Such a critic writes a bildungsroman about how English poetry came to be what it now is. Heidegger is writing a bildungsroman about, in his phrase, "what Being now is."[19] He tries to identify the philosophers, and the words, which have been decisive for getting Europe to the point where it now is. He wants to give us a genealogy of final vocabularies which will show us why we are currently using the final vocabulary we are by telling a story about the theorists (Heraclitus, Aristotle, Descartes, and so on), whom we have to go through rather than around. But the criterion of choice of figures to discuss, and of elementary words to isolate, is not that the philosophers or the words are authorities on something other than themselves – on, for instance, Being. They are not revealers of anything except us – us twentieth-century ironists. They reveal us because they *made* us. "The most elementary words in which Dasein expresses itself" are not "most elemental" in the sense that they are closer to how things are in themselves, but only in the sense that they are closer to *us*.

I can summarize my story about Heidegger by saying that he hoped to avoid Nietzsche's relapse from irony into metaphysics, his final surrender to his desire for power, by giving us a *litany* rather than a narrative. He thoroughly understood Hegel's and Nietzsche's problem of how to end their narrative, and – toward the end of his life – he hoped he had avoided the trap into which they fell by treating his narrative of the history of Being as merely a ladder which could be thrown away, merely an artifice for bringing the "elemental words" to our attention. He wanted to help us hear the words which had made us what we are. We were to do this, he finally decided, not for the sake of overcoming anything – for example, "Western ontology," or ourselves – but, rather

19 The difference is that a critic like Bloom remains distinct from the figure he constructs, whereas Heidegger tends to blend with it.

for the sake of *Gelassenheit,* of the ability *not* to seek power, the ability *not* to wish to overcome.[20]

The analogies between Heidegger's attempt, so described, and Proust's attempt, as I described it earlier, are fairly obvious. Proust's effort to deprive the concept of authority of authority, by redescribing all possible authorities as fellow sufferers, is paralleled by Heidegger's attempt simply to hear the resonances of the words of the metaphysicians rather than to use these words as instruments. His description of what he was doing as *andenkendes Denken* – a thinking that recalls – makes the analogy with Proust still easier. Both he and Proust thought that if memory could retrieve what created us, that retrieval itself would be tantamount to becoming what one was.

Having drawn this analogy, I can now explain what is wrong with it, why I think Heidegger failed where Proust succeeded. Proust succeeded because he had no public ambitions – no reason to believe that the sound of the name "Guermantes" would mean anything to anybody but his narrator. If that same name does in fact have resonance for lots of people nowadays, that is just because reading Proust's novel happens to have become, for those people, the same sort of thing which the walk *à côté de Guermantes* happened to become for Marcel – an experience which they need to redescribe, and thus to mesh with other experiences, if they are to succeed in their projects of self-creation. But Heidegger thought he knew some words which had, or should have had, resonance for *everybody* in modern Europe, words which were relevant not just to the fate of people who happen to have read a lot of philosophy books but to the public fate of the West. He was unable to believe that the words which meant so much to him – words like "Aristotle," *physis,* "Parmenides," *noien,* "Descartes," and *substantia* – were just his own private equivalents of "Guermantes," "Combray," and "Gilberte."

But that is, in fact, all they were. Heidegger was the greatest theoretical imagination of his time (outside the natural sciences); he achieved the sublimity he attempted. But this does not prevent his being entirely useless to people who do not share his associations. For people like me, who do share them, he is an exemplary, gigantic, unforgettable figure. Reading Heidegger has become one of the experiences with which we have to come to terms, to redescribe and make mesh with the rest of our experiences, in order to succeed in our own projects of self-creation. But Heidegger has no general public utility. To people who have never read, or read and were merely amused by, the attempts of metaphysicians like Plato and Kant to affiliate themselves with an

20 See the passage from "Time and Being" quoted in note 1, in this chapter.

ahistorical power, ironist theory seems an absurd overreaction to an empty threat. Such people will find Heidegger's *andenkendes Denken* no more urgent a project than Uncle Toby's attempt to construct a model of the fortifications of Namur.

Heidegger thought that he could, by virtue of his acquaintance with certain books, pick out certain words which stood to all contemporary Europeans as Marcel's litany of recollections stood to him. He could not. There is no such list of elementary words, no universal litany. The elementariness of elementary words, in Heidegger's sense of "elementary," is a private and idiosyncratic matter. The list of books which Heidegger read is no more central to Europe and its destiny than a lot of other lists of a lot of other books, and the concept of "the destiny of Europe" is, in any case, one we can do without. For this sort of historicist dramaturgy is just a further attempt to fend off thoughts of mortality with thoughts of affiliation and incarnation.[21]

Heidegger was quite right in saying that poetry shows what language can be when it is not a means to an end, but quite wrong in thinking that there could be a universal poem – something which combined the best features of philosophy and poetry, something which lay beyond both metaphysics and ironism. Phonemes *do* matter, but no one phoneme matters to very many people for very long. Heidegger's definition of "man" as "Being's poem" was a magnificent, but hopeless, attempt to save theory by poetizing it. But neither man in general nor Europe in particular has a destiny, nor does either stand to any larger-than-human figure as a poem stands to its author. Nor is ironist theory more than one of the great literary traditions of modern Europe – comparable to the modern novel in the greatness of the achievements which exemplify it, though far less relevant to politics, social hope, or human solidarity.

When Nietzsche and Heidegger stick to celebrating their personal canons, stick to the little things which meant most to them, they are as magnificent as Proust. They are figures whom the rest of us can use as examples and as material in our own attempts to create a new self by writing a bildungsroman about our old self. But as soon as either tries to

21 See Alan Megill, *Prophets of Extremity* (Berkeley: University of California Press, 1985), p. 346: "The notion of a crisis in history presupposes what it sets out to destroy – the idea of history as a continuous process, history with a capital H." Megill's criticism of what he calls Heidegger's "aestheticism" parallels my criticism of Heidegger's attempt at the historical sublime. Megill defines "aestheticism" as "an attempt to bring back into thought and into our lives that form of edification, that reawakening of *ekstasis,* which in the Enlightenment and post-Enlightenment view has been largely confined to the realm of art" (p. 342). In this sense of "aestheticism," my effort in this book (and especially in my sketch of a liberal utopia in Chapter 3) is to suggest that we bring this attempt into our private lives without trying to bring it into politics.

put forward a view about modern society, or the destiny of Europe, or contemporary politics, he becomes at best vapid, and at worst sadistic. When we read Heidegger as a philosophy professor who managed to transcend his own condition by using the names and the words of the great dead metaphysicians as elements of a personal litany, he is an immensely sympathetic figure. But as a philosopher of our public life, as a commentator on twentieth-century technology and politics, he is resentful, petty, squint-eyed, obsessive – and, at his occasional worst (as in his praise of Hitler after the Jews had been kicked out of the universities), cruel.

This claim repeats a suggestion I made at the end of the preceding chapter: that irony is of little public use, and that ironist theory is, if not exactly a contradiction in terms, at least so different from metaphysical theory as to be incapable of being judged in the same terms. Metaphysics hoped to bring together our private and our public lives by showing us that self-discovery and political utility could be united. It hoped to provide a final vocabulary which would not break apart into a private and a public portion. It hoped to be both beautiful on a small private scale and sublime on a large public one. Ironist theory ran its course in the attempt to achieve this same synthesis through narrative rather than system. But the attempt was hopeless.

Metaphysicians like Plato and Marx thought they could show that once philosophical theory had led us from appearance to reality we would be in a better position to be useful to our fellow human beings. They both hoped that the public-private split, the distinction between duty to self and duty to others, could be overcome. Marxism has been the envy of all later intellectual movements because it seemed, for a moment, to show how to synthesize self-creation and social responsibility, pagan heroism and Christian love, the detachment of the contemplative with the fervor of the revolutionary.

On my account of ironist culture, such opposites can be combined in a life but not synthesized in a theory. We should stop looking for a successor to Marxism, for a theory which fuses decency and sublimity. Ironists should reconcile themselves to a private-public split within their final vocabularies, to the fact that resolution of doubts about one's final vocabulary has nothing in particular to do with attempts to save other people from pain and humiliation. Colligation and redescription of the little things that are important to one – even if those little things are philosophy books – will not result in an understanding of anything larger than oneself, anything like "Europe" or "history." We should stop trying to combine self-creation and politics, especially if we are liberals. The part of a liberal ironist's final vocabulary which has to do with public action is never going to get subsumed under, or subsume, the rest of her final vocabulary. I shall claim in Chapter 8 that liberal political discourse would do well to remain as untheoretical and simpleminded as it looks (and as Orwell thought it), no matter how sophisticated the discourse of self-creation becomes.

6

From ironist theory to private allusions: Derrida

Derrida stands to Heidegger as Heidegger to Nietzsche. Each is the most intelligent reader, and most devastating critic, of his respective predecessor. That predecessor is the person from whom each has learned most, and whom he most needs to surpass. Derrida continues to think about the problem which came to obsess Heidegger: that of how to combine irony and theorizing. But he has the advantage of having observed Heidegger's failure, as Nietzsche and Heidegger had the advantage of having observed Hegel's.

Derrida learns from Heidegger that phonemes matter, but he realizes that Heidegger's litany is just Heidegger's, not Being's or Europe's. His problem becomes, as he says at the end of "Différance," to think the fact that "there is no unique name" – or, more generally, no definitive litany – "without *nostalgia*, that is, outside of the myth of a purely maternal or paternal language, a lost native country of thought."[1] He wants to figure out how to break with the temptation to identify himself with something big – something like "Europe" or the "call of Being" or "man." As he says in his reply to Heidegger's *Letter on Humanism*, Heidegger's use of "we" arises out of the "eschatoteleological situation" which is "inscribed in metaphysics."[2] The trouble with Heidegger's thinking of "the presence of the present" is that it "can only metaphorize, by means of a profound necessity from which one cannot simply decide to escape, the language that it deconstructs."[3] Heidegger's *andenkendes Denken* is nostalgic or nothing, and the myth of a lost language, of "elementary words" whose force needs to be restored, is just one more attempt to believe that some words are privileged over others by a power not ourselves, that some final vocabularies are closer to something transhistorical and noncontingent than others.[4]

---

1 Jacques Derrida, *Margins of Philosophy* (Chicago: University of Chicago Press, 1982), p. 27.
2 Ibid., p. 123.
3 Ibid., p. 131.
4 Much of my criticism of Heidegger in Chapter 5 is borrowed from Derrida, and in particular from "The Ends of Man" and from "Différance." For a very acute characterization and criticism of the reading of Heidegger which I derive from Derrida, see John D. Caputo, "The Thought of Being and the Conversation of Mankind: The Case

Like Heidegger's, Derrida's work divides into an earlier, more professorial period and a later period in which his writing becomes more eccentric, personal, and original. In *Being and Time*, as I said earlier, Heidegger pours Nietzschean wine into Kantian vessels. He says Nietzschean things in the context of the standard German academic project of finding "conditions of the possibility" of familiar experiences. Derrida's earlier work can also be read as such a project – the project of going deeper than Heidegger went, in quest of the same sort of thing Heidegger wanted: words which express the conditions of possibility of all previous theory – all of metaphysics and all earlier attempts to undercut metaphysics, including Heidegger's. On this reading, Derrida wants to undercut Heidegger as Heidegger undercut Nietzsche. Yet his project is continuous with Heidegger's in that he, too, wants to find words which get us "beyond" metaphysics – words which have force apart from us and display their own contingency.

Many of Derrida's admirers, notably Rodolphe Gasché, read his earlier work in this way. But Gasché begins his book by saying that he will not discuss *Glas* or Derrida's work after *The Truth in Painting*, and that he puts aside "the delicate question of what is to be counted as more philosophical or more literarily playful."[5] Gasché proceeds to reconstruct Derrida's early work as an attempt to formulate a "system beyond Being," a system of "infrastructures" (e.g., *différance,* spacing, iterability)

of Heidegger and Rorty," *Review of Metaphysics* 36 (1983): 661–685. Caputo is right in saying that, like Derrida, I am "interested in the destruction of the history of ontology in its negative sense" and think the idea that it has a "positive sense" is "Heidegger's final illusion" (p. 676). But he is wrong in saying that my view, or Derrida's, ensures that "we get no further than propositional discourse" (pp. 677–678). All that I (or, as far as I can see, Derrida) want to exclude is the attempt to be nonpropositional (poetic, world-disclosing) and at the same time claim that one is getting down to something primordial – what Caputo calls "the silence from which all language springs" (p. 675). The nominalism described in Chapter 1 (a nominalism I take Derrida to share) demands that we reject Caputo's Heideggerian claim that "language is not a system of words devised for human purposes but the event which gives birth to things." That claim seems to me a confusion of causal conditions with the mysterious transcendental "conditions of possibility" dreamed up by Kant. As I said at the beginning of Chapter 1, we nominalists want to cleanse Romanticism of the last traces of German idealism – and, as I argue below, this means eliminating *argumentative appeal* to the nonpropositional. This is precisely what Derrida, Davidson, and Bloom help us to do, by letting us think of poets as themselves *ursprünglich* rather than passive recipients of the gifts of Being. By contrast, as Caputo says, for Heidegger "the authentic speaker . . . is taken over by the things themselves, yields to them, lets them come to words in him" (p. 674). This wish for association with Something Other and Bigger is just what Derrida distrusts most in Heidegger, and I think he is right to do so. For a good reply to Caputo's article, see Lyell Asher, "Heidegger, Rorty and the Possibility of Being" in *Ethics/Aesthetics: Post-Modern Positions,* ed. Robert Merrill (Washington, D.C.: Maisonneuve Press, 1988).

5 Rodolphe Gasché, *The Tain of the Mirror: Derrida and the Philosophy of Reflection* (Cambridge, Mass.: Harvard University Press, 1986), p. 4.

which take us behind, or beneath, Heidegger.[6] He takes Derrida to have "demonstrated" that

. . . the "source" of all being beyond being is *generalized,* or rather *general,* writing, whose essential nontruth and nonpresence is the fundamentally undecidable condition of possibility and impossibility of presence in its identity and of identity in its presence. The "source" of being and beingness is, for Derrida, the system or chain beyond being of the various infrastructures or undecidables.[7]

There is much in Derrida's early work which encourages this reading, and I shall not take up the question of the accuracy of Gasché's description to Derrida's early intentions. But there is an obvious problem with any such reading, namely, that the whole idea of "undercutting" and of "conditions of possibility" sounds terribly metaphysical. That is, it seems to presuppose that there is a fixed vocabulary within which such a project can be carried out – that all the people whom Gasché calls "philosophers of reflection" know what it is like to find a "condition of possibility," and can tell who has succeeded in undercutting whom.[8]

6　Gasché takes seriously Derrida's claim that "différance" is "neither a word nor a concept" and applies it to all the other Derridean terms which he takes to signify infrastructures. I have criticized this claim in "Deconstruction and Circumvention," *Critical Inquiry* 11 (1983): 1–23, arguing that here Derrida tries, like Heidegger, to have it both ways: to eff the ineffable by decreeing that a word which he puts into circulation is not the sort of thing which can be put into circulation. Like late Heidegger, early Derrida sometimes goes in for word magic – hoping to find a word which cannot be banalized and metaphysicized by being used, which will somehow retain its "instability" even after it becomes current. Gasché seems to think that this magic works, as when he says, "An infrastructure, moreover, is not an essence, since it is not dependent on any category of that which is present or absent. . . . It has no stable character, no autonomy, no ideal identity, and thus is not a *substance,* or *hypokeimenon.* Its 'essence' is to have no essence. And yet an infrastructure is endowed with a certain universality." Paying these compliments to Derrida's words seems to me whistling in the dark – saying that it would be nice if there were words which had this impossible combination of properties without explaining how the combination is supposed to have been made possible. The superiority of later to earlier Derrida seems to me precisely that he stops relying on word magic and relies instead on a way of writing – on creating a style rather than on inventing neologisms.

7　Gasché, *The Tain of the Mirror,* p. 177.

8　I discuss Gasché's book in some detail in "!s Derrida a Transcendental Philosopher?" *The Yale Journal of Criticism* (in press). See also an exchange between Christopher Norris and myself on the question of whether Derrida should be regarded as "playful" or "serious" in *Redrawing the Lines: Analytic Philosophy, Deconstruction and Literary Theory,* ed. Reed Dasenbrock (Minneapolis: University of Minnesota Press, in press). Norris's contribution – "Philosophy as *Not* Just a 'Kind of Writing': Derrida and the Claim of Reason" – is in part a reply to my "Philosophy as a Kind of Writing: An Essay on Derrida," included in my *Consequences of Pragmatism* (Minneapolis: University of Minnesota Press, 1982). My contribution – "Two Senses of 'Logocentrism': A Reply to Norris" – argues against the claim that Derrida gives philosophical foundations for so-called deconstructive literary criticism, and also argues that Derrida's outlook and strategy differ dramatically from those of Paul de Man, the writer whose work set the tone for that kind of criticism.

Gasché's description of the search for "conditions of the possibility of philosophical discourse" suggests that understanding what such conditions are, and knowing how to search for such conditions, is a matter of what Gasché calls the "standard rules of philosophy."[9] But it would be odd if reference to such rules enabled us to shelve the problem I mentioned earlier in discussing Heidegger: that the realm of possibility expands whenever a new vocabulary is invented, so that to find "conditions of possibility" would require us to envisage all such inventions before their occurrence. The idea that we do have such a metavocabulary at our disposal, one which gives us a "logical space" in which to "place" anything which anybody will ever say, seems just one more version of the dream of "presence" from which ironists since Hegel have been trying to wake us.

Whether or not Derrida was initially tempted by the transcendental project which Gasché ascribes to him, I suggest that we read Derrida's later writings as turning such systematic projects of undercutting into private jokes. In my view, Derrida's eventual solution to the problem of how to avoid the Heideggerian "we," and, more generally, avoid the trap into which Heidegger fell by attempting to affiliate with or incarnate something larger than himself, consists in what Gasché refers to disdainfully as "wild and private lucubrations."[10] The later Derrida privatizes his philosophical thinking, and thereby breaks down the tension between ironism and theorizing. He simply drops theory – the attempt to see his predecessors steadily and whole – in favor of fantasizing about those predecessors, playing with them, giving free rein to the trains of associations they produce. There is no moral to these fantasies, nor any public (pedagogic or political) use to be made of them; but, for Derrida's readers, they may nevertheless be exemplary – suggestions of the sort of thing one might do, a sort of thing rarely done before.

Such fantasizing is, in my view, the end product of ironist theorizing. Falling back on private fantasy is the only solution to the self-referential problem which such theorizing encounters, the problem of how to distance one's predecessors without doing exactly what one has repudiated them for doing. So I take Derrida's importance to lie in his having had the courage to give up the attempt to unite the private and the public, to stop trying to bring together a quest for private autonomy and an attempt at public resonance and utility. He privatizes the sublime, having learned from the fate of his predecessors that the public can never be more than beautiful.

9 Gasché, *Tain,* p. 122.
10 Ibid., p. 123.

The quest for the sublime was, in Heidegger, the quest for words which had "force," rather than the mere exchange value given them by their role in language games. The dilemma Heidegger faced was that as soon as he isolated such words and published his results, the words promptly became part of the widely played Heideggerian language game, and were thereby demoted from *Winke* to *Zeichen,* from Thinking to metaphysics. As soon as he went public, his "elementary words" lost force by gaining uses (e.g., becoming names of "philosophical problems" – the "problem of presence," the "problem of technology," etc.). Derrida learned from Heidegger's example that the problem is not "to touch upon the nature of language without doing it injury" but rather to create a style so different as to make one's books incommensurable with those of one's precursors. He learned that "language" no more has a nature than "Being" or "man" does, and that the attempt to pare language down to "elementary words" was futile.

So, instead of paring down, the later Derrida proliferates. Instead of hoping, with Heidegger, always to "say the same," "to bring to language ever and again this advent of Being which remains . . . the sole matter of thinking,"[11] he takes pains never to say the same thing twice. Whereas in Heidegger you know that whatever the purported topic of the essay, you will come back around to the need to distinguish beings from Being, or to remember Being, or to be grateful to Being, in later Derrida you never know what is coming next. Derrida is interested not in the "splendor of the simple" but, rather, in the lubriciousness of the tangled. He is interested neither in purity nor in ineffability. *All* that connects him with the philosophical tradition is that past philosophers are the topics of his most vivid fantasies.

The first half of *The Post Card,* entitled "Envois," is the text which best illustrates what I take to be Derrida at his best. "Envois" differs from *Glas* in being readable, and also in being moving. It owes these features to its form – a sequence of love letters. This form emphasizes the privacy of the work being done. Nothing is more private than a love letter – there is nothing to which general ideas are less relevant or more inappropriate. Everything, in a love affair or a love letter, depends upon shared private associations, as when the "traveling salesman"[12] who

---

11 Heidegger, "Letter on Humanism," in *Basic Writings,* ed. David Krell (New York: Harper & Row, 1977), p. 241.

12 "I write you the letters of a traveling salesman hoping that you hear the laughter and the song – the only ones (the only what?) that cannot be sent, nor the tears. At bottom I am only interested in what cannot be sent off, cannot be dispatched in any case" (*The Post Card from Socrates to Freud and Beyond,* trans. Alan Bass [Chicago, University of Chicago Press, 1987], p. 14). The original is at *La Carte Postale de Socrate à Freud et au*

writes the letters in "Envois" recalls "the day when we bought that bed (the complications of credit and of the perforated tag in the department store, and then one of those horrible scenes between us)."[13] The letters get much of their poignancy from the references to real-life events and people: landing at Heathrow, lecturing at Oxford, landing at Kennedy, teaching at Yale, recovering from an accident with a skateboard, talking across the ocean by phone ("and then you laugh and the Atlantic recedes").[14]

In the course of the letters, the author spins out fantasies about a postcard he has come across in Oxford: a reproduction of a thirteenth-century picture showing two figures – one labeled "plato" and the other "Socrates." He writes his love letters on the backs of innumerable copies of this postcard, and fantasizes endlessly about the relations between Socrates and Plato. This pair eventually gets run together with lots of other pairs: Freud and Heidegger, Derrida's two grandfathers, Heidegger and Being, Beings and Being, Subject and Object, S and p, the writer himself and "you," his "sweet love" – and even "Fido" and Fido.[15] The fantasies are, like the letters themselves, a mixture of the privately erotic and the publicly philosophical. They mingle idiosyncratic obsessions with reflections on the paradigmatic attempt to escape from the merely private – metaphysics, the search for generality.

The usual picture of Socrates is of an ugly little plebeian who inspired a handsome young nobleman to write long dialogues on large topics. Perhaps because some copyist put the wrong names next to the figures in

---

delà (Paris: Aubier-Flammarion, 1980), p. 19. (Henceforth I shall give page numbers from both the translation and the original, with that of the translation first.)

13   Ibid., p. 34/40.

14   However, Derrida keeps us guessing about whether all the letters are written by the same person, or are addressed to the same person, and also about whether the "sweet love" (or loves) in question is (or are) male or female, real or imaginary, concrete or abstract, identical or different from the writer (or from you, the reader of the book), and so on. At p. 5/9 he says: "That the signers and the addressees are not always visibly and necessarily identical from one *envoi* to the other, that the signers are not inevitably to be confused with the senders, nor the addressees with the receivers, that is with the readers (*you* for example), etc. – you will have the experience of all this, and sometimes will feel it quite vividly, although confusedly."

15   "S" and "P" – for "subject" and "predicate," respectively – are abbreviations frequently found in works of analytic philosophy. (But on the postcard in question, "plato" is written with a small "p" and "Socrates" with a large "S," so Derrida decapitalizes "p" throughout.) We Derrida admirers are tempted to write learnedly on the relation between the S-p relation in "Envois" and the S-a relation ("Savoir absolu," Lacan's "petit a," and all that) in *Glas* – but such temptations should be resisted. Nobody wants a complete set of footnotes to *The Post Card* any more than they want one to *Finnegans Wake, Tristram Shandy,* or *Remembrance of Things Past.* The reader's relation with the authors of such books depends largely upon her being left alone to dream up her own footnotes.

the picture, the postcard shows "plato" as an ugly, ill-dressed little man standing behind, and noodging, a big, well-dressed "Socrates." The latter is seated at a desk, busy writing something. For no clear reason, there is a big something (looking a bit like a skateboard) sticking out from between Socrates' rear end and the chair he is sitting in – a something Derrida promptly interprets as obscenely as possible:

> For the moment, myself, I tell you that I see *Plato* getting an erection in *Socrates'* back and see the insane hubris of his prick, an interminable, disproportionate erection traversing Paris' head like a single idea and then the copyist's chair, before slowly sliding, still warm, under *Socrates'* right leg, in harmony or symphony with the movement of this phallus sheaf [*ce faisceau de phallus*], the points, plumes, pens, fingers, nails and *grattoirs,* the very pencil boxes which address themselves in the same direction.[16]

From here on, Derrida rings every possible change on influencing philosophers, creating fictional philosophers to serve as dummies, standing philosophers on their heads, penetrating them from the rear, fertilizing them so that they give birth to new ideas, and so on. More and more associations cluster around, and eventually three names appear more and more often – those of Freud (who concentrated on sexual inversions and misdirections), of Heidegger (who concentrated on dialectical inversions and misreadings), and of Fido (about whom more later).

As one example of the sort of thing Derrida gives us in "Envois," consider his conflation of the metaphysical urge for a privileged final vocabulary, for general ideas, with the urge to have children (an echo of Socrates' talk of "midwifery" and "wind-eggs" in the *Theaetetus*). Early in the correspondence he says to his "sweet love" that "what has betrayed us, is that you wanted generality, which is what I call a child."[17] Children, like the universal public truths (or privileged descriptions, or unique names) which metaphysicians hope to hand down to posterity, are traditionally thought to be a way of evading death and finitude. But children, and the succeeding generation of philosophers, tend to patricide and matricide – a fact which leads Derrida to write, "At least help me so that death comes to us only from us. Do not give in to generality."[18] Further, it is hard to tell who the parents of a child or a philosophy are. In the letter, after he has introduced the theme of children, Derrida writes,

---

16 Ibid., p. 18/22–23. "Paris" is a reference to Matthew of Paris, the author of the book on fortune-telling which the picture on the postcard illustrates. The obscenity of the scene Derrida envisages contrasts with the chastity which Plato imputes to Socrates (his refusal to have sex with Alcibiades, and perhaps also with Plato himself).

17 P. 23/28.

18 P. 118/130.

"The two imposters' [plato's and Socrates'] program is to have a child by me, them too."

Immediately thereafter, however, he writes:

To the devil with the child, the only thing we ever will have discussed, the child, the child, the child. The impossible message between us. A child is what one should not be able to "send" oneself. It never will be, never *should* be a sign, a letter, even a symbol. Writings: stillborn children one sends oneself in order to stop hearing about them – precisely because children are first of all what one wishes to hear speak by themselves. Or this is what the two old men say.[19]

In Derrida's view, nothing ever speaks "by itself," because nothing has the primordiality – the nonrelational, absolute, character – metaphysicians seek. Nevertheless, we cannot help wanting to produce something which will so speak. If there *were* a "unique name," an "elementary word," or "conditionless condition of possibility," this would be, for Derrida, a tragedy: "For the day that there will be a reading of the Oxford card, the one and true reading, will be the end of history. Or the becoming-prose of our love."[20] If I am right in thinking of Derrida's later manner as a rejection of the transcendental temptations of his earlier one, then we can take the claim "what I will never resign myself to is publishing anything other than post cards, speaking to *them*"[21] as saying "I shall send you no children, just post cards, no public generalities, just private idiosyncrasies."[22]

The incredible richness of texture of "Envois" – a richness achieved by few other contemporary writers, and no other contemporary philosophy professors – is nicely illustrated by this playing off of one's feelings about babies against one's feelings about books. It is the sort of assimilation which links up with much else in Derrida – for example, the contrast in *On Grammatology* between the (interminable) text and the (terminable) book, which now appears as the contrast between love for its own sake and love for the sake of making babies. Somebody who writes nothing but postcards will not have Hegel's problem of how to end his book, nor

19 P. 25/29–30. Compare this passage with the end of Heidegger's "Time and Being," cited in the preceding chapter: "Yet a regard for metaphysics still prevails in the intention to overcome metaphysics. Therefore our task is to cease all overcoming, and leave metaphysics to itself." Imagine Heidegger saying, "To the devil with metaphysics, the only thing I will ever have discussed."

20 P. 115/127.

21 P. 13/17. I take "them" to be plato and Socrates.

22 But Derrida's constant preoccupation with the self-referential paradox involved in his making any general programmatic statement is illustrated by p. 238/255: "They will never know if I do or do not love the post card, if I am for or against."

Gasché's problem of knowing whether he has hit rock bottom in the search for infrastructures. But he will also not produce a "result," a "conclusion." There will be no "upshot" – nothing to carry away from "Envois" (lovingly, cupped in one's hands or cradled in one's arms) once one has finished reading it.

This reduction of public to private productions, of books to babies, writing to sex, thinking to love, the desire for Hegelian absolute knowledge to the desire for a child,[23] is continued when Derrida conflates Freud and Heidegger:

Here Freud and Heidegger, I conjoin them within me like the two great ghosts of the "great epoch." The two surviving grandfathers. They did not know each other, but according to me they form a couple, and in fact just because of that, this singular anachrony. They are bound to each other without reading each other and without corresponding. I have often spoken to you about this situation, and it is this picture that I would like to describe in *Le legs:* two thinkers whose glances never crossed and who, without ever receiving a word from each other, say the same. They are turned to the same side.[24]

---

23 See p. 39/44–45: "the child remaining, alive or dead, the most beautiful and most living of fantasies, as extravagant as absolute knowledge."

24 P. 191/206. The reference to *Le legs* is to *"Legs de Freud,"* one of the essays included in the second half of *The Post Card.* There Derrida discusses, among other things, Freud's children (especially Sophie and Ernst); the title is an Anglo-French pun, as well as being ambiguous between Freud's books and Freud's babies. The reference to the "great epoch" [*la grande époque*] is to "the great epoch (whose technology is marked by paper, pen, the envelope, the individual subject addressee, etc.) and which goes shall we say from Socrates to Freud and Heidegger" (ibid.). This is the epoch which, in Derrida's earlier jargon, is that of the "book" (as opposed, initially, to the text and later to the postcard). It is also the epoch which Heidegger identified as that of "Western metaphysics" – the "logocentric" epoch which centered around the search for what Husserl called an "epoche": grasping essence through decontextualization. "Say the same" is an ironic reference to Heidegger's use of that phrase. In the postcard of plato and Socrates, the two are turned to the same side, and their glances do not meet. On grandfathers, compare p. 61/68, where Derrida describes Socrates (on the postcard) as "young, as is said in [Plato's Second] Letter, younger than *Plato,* and handsomer, and bigger, his big son, his grandfather or his big grandson, his *grandson.*" Derrida says in this passage, "It is S. [the subject of Plato's dialogues – i.e., Socrates] who has written everything" that Plato wrote, a claim which refers back to the fact that "Plato's dream" was "to make Socrates write, and to make him write what he wants, his last command, *his will* . . . thereby becoming Socrates and his father, therefore his own grandfather" (p. 52/59). The reference is also to a passage in the Second Letter where Plato says, "There is not and will not be any written work of Plato's own. What are now called his are the work of a Socrates grown young and beautiful."
    The length of this (sternly curtailed) footnote may suggest to those unfamiliar with "Envois" what I mean by its "richness of texture" – a richness in part made possible by taking "mere associations" between noises and marks seriously. Almost any half-dozen lines from "Envois" could beget an equally lengthy footnote.

What is this "same" which Heidegger and Freud – the specialist in Being and the detector of dirty little secrets – say? They can be interpreted as saying a *lot* of the same things, so perhaps the question is "Why does Derrida think that this *particular* couple marks the end of a great epoch which begins with the coupling of Plato and Socrates?" The best answer I can think of is that both Heidegger and Freud were willing to attach significance to phonemes and graphemes – to the shapes and sounds of words. In Freud's account of the unconscious origins of jokes, and in Heidegger's (largely fake) etymologies, we get the same attention to what most of the books of *la grande époque* have treated as inessential – the "material" and "accidental" features of the marks and noises people use to get what they want. If this answer is at least partly right, then the constant recourse to puns, verbal resonances, and graphical jokes in Derrida's later work is what we should expect of somebody who has resolved to "send only postcards." For the only way to evade problems about how to end books, to escape self-referential criticisms that one has done what one has accused others of doing, will be to transfer the weight of one's writing to those "material" features – to what has hitherto been treated as marginal. These associations are necessarily private; for insofar as they become public they find their way into dictionaries and encyclopedias.[25]

This brings us to another couple: "Fido"-Fido. Reference to this one are about as frequent in "Envois" as to Freud-Heidegger. "Fido" is the name of the dog Fido, just as ""Fido"" is the name of the name of that dog. (Notice that new names, but not new dogs, can be produced as fast as one can stick in quotation marks.) Oxford philosophers (e.g., P.H. Nowell-Smith, Gilbert Ryle) dubbed the idea that "all words are names" the ""Fido"-Fido theory of meaning." This theory was often associated (by Austin, among others) with Plato. It contrasts with the view, associated with Saussure and Wittgenstein, that words get their sense not simply by association with their referents (if any) but by the relation of their uses to the uses of other words. (You might, given some stage-setting, learn the use of "Fido" by somebody pointing to Fido and saying, "That's Fido," but you do not learn the use of "good" by dimly recollecting the Form thereof and labeling this memory with that vocable, nor of "I" by labeling a salient feature of yourself.)

The second occasion on which "Fido"-Fido comes up in "Envois" is in the course of an enormous multipostcard postscript about "an in-

---

25 For example, the association between "Hegel" and "Hegelian" or Hegel and Spirit, is public. Derrida's association between "Hegel" and *aigle* is private.

complete pair of terrible grandfathers . . . the couple Plato/Socrates, divisible and indivisible, their interminable partition, the contract which binds them to us until the end of time." There Derrida says:

This is the problem of "'Fido'-Fido" (you know, Ryle, Russell, etc.), and the question of knowing whether I am calling my dog or if I am mentioning the name of which he is the bearer, if I am utilizing or if I am naming his name. I adore these theorizations, often Oxonian moreover, their extraordinary and necessary subtlety as much as their imperturbable ingenuity, *psychoanalytically speaking;* they will always be confident in the law of quotation marks.[26]

The difference between "Fido" and Fido is often used to illustrate Russell's distinction between "mentioning" a word (in order to say, e.g., that it has four letters) and "using" it (in order, e.g., to call a dog). This distinction enables us to separate the "essential" meaning, or use, or function, of "Fido" and the "accidental" features of this name (e.g., that, qua mark or noise, it reminds one of the Latin verb *fidere* and thus of fidelity, and thus of characters in literature such as Dudley Doright's faithful dog Faithful Dog, and so on and on).[27] Years ago, John Searle, "confident in the law of quotation marks," charged Derrida with having neglected this distinction when discussing Austin's work.[28] Derrida replied by raising doubts about the utility and scope of the distinction itself − doubts which were exasperatingly irrelevant to Searle's complaints. For Searle was saying, "If you play by the rules of Austin's language game, if you respect his motives and intentions, your criticisms of him do not work. If, on the other hand, you feel free to read whatever you like into him, if you treat him psychoanalytically, for example, you cannot claim to be *criticizing* him; you are simply using him as a figure in your own fantasies, giving free rein to a set of associations which have no relevance to Austin's project."

In his reply to Searle,[29] Derrida systematically evaded this dilemma.

26 P. 98/108.
27 See p. 243/260−61: "Ah yes, Fido, I am faithful to you as a dog. Why did 'Ryle' choose this name, Fido? Because one says of a dog that he *answers* to his name, to the name of Fido, for example? Because a dog is the figure of fidelity and that better than anyone else answers to his name, especially if it is Fido? . . . Why did Ryle choose a dog's name, Fido? I have just spoken at length about this with Pierre, who whispers to me: 'so that the example will be obedient.'" Notice the disregard of the use-mention distinction in this passage. Note also a neighboring passage about Anglo-Saxon philosophers: "But there always comes a moment when I see their anger mount on a common front; their resistance is unanimous: 'and quotation marks − they are not to go to the dogs! [*les guillemets, c'est pas pour les chiens!*] and theory, and meaning, and reference, and language!' *Mais si, mais si.*"
28 See Searle, "Reiterating the Differences: A Reply to Derrida," *Glyph* 1 (1977): 198−208.
29 Derrida, "Limited Inc," *Glyph* 2 (1977): 162−254.

But why did he not just grasp its second horn? Why was his reply to Searle twice as fantastical and free-associative as his original criticism of Austin, while nevertheless being filled with straight-faced professions of sincerity and seriousness? Presumably for the same reason that he would resist the Gasché-like question "Is 'Envois' to be counted as philosophical or as literary and playful?" I take it that Derrida does not want to make a single move within the language game which distinguishes between fantasy and argument, philosophy and literature, serious writing and playful writing – the language game of *la grande époque*. He is not going to play by the rules of somebody else's final vocabulary.

He refuses not because he is "irrational," or "lost in fantasy," or too dumb to understand what Austin and Searle are up to but because he is trying to create himself by creating his own language game, trying to avoid bearing another child by Socrates, being another footnote to Plato. He is trying to get a game going which cuts right across the rational-irrational distinction. But, as a philosophy professor, he has trouble getting away with this.[30] Whereas it would be pretty crude to ask Proust whether we should read his novel as social history or as a study of sexual obsession, or to ask Yeats whether he really *believed* all that guff about phases of the moon, philosophers are traditionally supposed to answer this sort of question. If you advertise yourself as a novelist or a poet you are let off a lot of bad questions, because of the numinous haze that surrounds the "creative artist." But philosophy professors are supposed to be made of sterner stuff and to stay out in the open.

This haze surrounds writers who are not associated with any particular discipline, and are therefore not expected to play by any antecedently known rules.[31] I have been urging that we enfold Derrida in this nimbus by seeing his purpose as the same autonomy at which Proust and Yeats aimed. The advantage of doing so is that we can avoid dissecting his writing along lines laid down by somebody else, and can instead sit back and enjoy it – wait to see what comfort or example it might offer us, whether it turns out to be relevant to our own attempts at autonomy. If we have not been impressed by Plato or Heidegger, the chances are it will be no use at all; if we have, it might be decisive. Anybody who has

30 I suspect that if Derrida were a rich belletrist who, having started off in poetry and fiction, had switched to philosophy, but had never had to earn his bread by teaching it, he would not get nearly as hard a time as he does from his professional colleagues.

31 As Jonathan Culler says, "The idea of a discipline is the idea of an investigation in which writing might be brought to an end" (*On Deconstruction* [Ithaca, N.Y.: Cornell University Press, 1982], p. 90). A writer who prides himself on his facility at proliferating loose ends is not going to contribute to a discipline, but that does not mean that he is undisciplined. A private discipline is not a discipline in Culler's "public" sense of the term, but it may nevertheless entail a lot of hard and exacting work.

read little of philosophy will get little from "Envois," but for a certain small audience it may be a very important book.

Accepting this suggestion means giving up the attempt to say, with Gasché and Culler, that Derrida has demonstrated anything or refuted anybody (e.g., Austin). It also means giving up the idea that Derrida has developed a "deconstructive method" which "rigorously" shows how the "higher" of a pair of opposed concepts (e.g., form-matter, presence-absence, one-many, master-slave, French-American, Fido-"Fido") "deconstructs itself." Concepts do not kill anything, even themselves; people kill concepts. It took Hegel a lot of hard work to manage the dialectical inversions he then pretended to have observed rather than produced. It takes a lot of hard work to produce such special effects as "presence is just a special case of absence" or "use is but a special case of mentioning."[32] Nothing except lack of ingenuity stands in the way of any such recontextualization, but there is no *method* involved, if a method is a procedure which can be taught by reference to rules.[33] Deconstruction is not a novel procedure made possible by a recent philosophical discovery. Recontextualization in general, and inverting hierarchies in particular, has been going on for a long time. Socrates recontextualized Homer; Augustine recontextualized the pagan virtues, turning them into splendid vices, and then Nietzsche reinverted the hierarchy; Hegel recontextualized Socrates and Augustine in order to make both into equally *aufgehoben* predecessors; Proust recontextualized (over and over again) everybody he met; and Derrida recontextualizes (over and over again) Hegel, Austin, Searle, and everybody else he reads.

But why does it sound so shockingly different when Derrida does it, if it is just dialectical inversion all over again? Simply because Derrida makes use of the "accidental" material features of words, whereas Hegel, though refusing to play by the rule that the "contradiction" relation can hold only between propositions and not between concepts, still stuck to the rule that you cannot put any weight on words' sounds and shapes.[34]

---

32 The last example is Culler's. He says, "Derrida is quite right to claim that use/mention is ultimately a hierarchy of the same sort as serious/nonserious and speech/ writing. All attempt to control language by characterizing distinctive aspects of its iterability as parasitic or derivative. A deconstructive reading would demonstrate that the hierarchy should be reversed and that *use* is but a special case of *mentioning*" (*On Deconstruction*, p. 120n).

33 One learns to "deconstruct texts" in the same way in which one learns to detect sexual imagery, or bourgeois ideology, or seven types of ambiguity in texts; it is like learning how to ride a bicycle or play the flute. Some people have a knack for it, and others will always be rather clumsy at it – but doing it is not facilitated or hindered by "philosophical discoveries" about, for example, the nature of language, any more than bicycle riding is helped or hindered by discoveries about the nature of energy.

34 The *Phenomenology* was pretty shockingly autonomous, too, in its day – the days before

Derrida's attitude toward all such rules is that it is, of course, necessary to follow them if one wants to argue with other people, but that there are other things to do with philosophers than argue with them.[35] These rules make argumentative discourse possible, but Derrida answers the question "What would happen if we ignored them?" His answer is given *ambulando* as Derrida writes the kind of prose we find in *Glas* and "Envois" – the kind in which you can never tell, from line to line, whether you are supposed to pay attention to the "symbolic" or the "material" features of the words being used. When reading *Glas* or "Envois," you quickly lose interest in the question "Should I view this thing qua signifier, or qua mark?" For purposes of reading this sort of text, the use-mention distinction is just a distraction.[36]

What is the good of writing that way? If one wants arguments which reach conclusions, it is no good at all. As I have said already, there is nothing propositional to be taken away from the experience of reading it – any more than from the writings of the later Heidegger. So is it to be judged by "literary" rather than "philosophical" criteria? No, because, as in the cases of the *Phenomenology of Spirit, Remembrance of Things Past,* and *Finnegans Wake,* there are no antecedently available criteria of *either* sort. The more original a book or a kind of writing is, the more unprecedented, the less likely we are to have criteria in hand, and the less point there is in trying to assign it to a genre. We have to see whether we can find a use for it. If we can, then there will be time enough to stretch the borders of some genre or other far enough to slip it in, and to draw up criteria according to which it is a good kind of writing to have invented. Only metaphysicians think that our present genres and criteria exhaust the realm of possibility. Ironists continue to expand that realm.

Why might one, nevertheless, be inclined to say that "Envois" counts as "philosophy," even though it does not issue in anything that could conceivably be called a philosophical *theory*? Well, for one thing, because

Hegel became a great dead philosopher. Hegel, too, had admirers (e.g., Engels and Lenin) who believed that he had discovered a "method," just as Culler and others believe that Derrida has discovered one.

35 Consider another of Derrida's responses to Searle: "Yeah, okay, nothing to be said against the *laws* which govern this problematic [of use and mention], if not to ask the question of the law, and of the law of the proper name as concerns those pairs called quotation marks. I say (to them and to you, my beloved) this is my body, at work, love me, analyze the corpus that I tender to you, that I extend here on this bed of paper, sort out the quotation marks from the hairs, from head to toe, and if you love me enough you will send me some news. Then you will bury me in order to sleep peacefully. You will forget me, me and my name" (p. 99/109).

36 See p. 186/201: "You will never get to know, nor will they, whether, when I use a name it is in order to say *Socrates* is me or 'Socrates' has seven letters. This is why one never will be able to translate."

only people who habitually read philosophy could possibly enjoy it. Might one also say that it has philosophy as one of its causes and one of its topics? Not exactly. It would be better to say that it counts *philosophers* – particular philosophers – among its causes and topics. More and more, as Derrida goes along, his relation is not to the doctrines of Plato or Heidegger, but to the men themselves.[37] Whereas ironist theorizing from Hegel to Heidegger was about metaphysical theorizing, Derrida's early writing was about ironist theorizing. As their writing was in danger of turning into more metaphysics, his was in danger of turning, if not into metaphysics, at least into more theorizing. His later writing avoids this danger, partly because it is about theori*zers*. Referring back to the opposition I sketched in Chapter 5, Derrida is coming to resemble Nietzsche less and less and Proust more and more. He is concerned less and less with the sublime and ineffable, and more and more with the beautiful, if fantastical, rearrangement of what he remembers.

I said in Chapter 5 that Proust reacted not to general ideas but to the people he knew as a child (e.g., his grandmother) or happened to meet later on (e.g., Charles Haas, Mme. Greffulhe, Robert de Montesquieu). Analogously, Derrida reacts to the grandfathers and housekeepers upon whose knees philosophy professors are dandled (e.g., Plato and Socrates, Windelband and Wilamowitz) and to people whom he has bumped into in the course of his career (e.g., Austin, Matthew of Paris, Searle, Ryle, Fido).[38] I also said that Proust's triumph was to have written a book which evaded any of the descriptions which the authority figures of his acquaintance had applied to him (or which he imagined they would have applied to him). Proust wrote a new kind of book; nobody had ever *thought* of something like *Remembrance of Things Past*. Now, of course, we all think of it – or, at least, anybody who wants to write a bildungs-

---

37 "Martin [Heidegger] has the face of an old Jew from Algiers" (p. 189/204); "The Geviert too, the loveliest post card that Martin has sent us from Freiburg . . ." (p. 67/75).

38 Fido is not exactly a person, and not exactly a dog or a name either. Still, by the end of "Envois" the reader feels on reasonably familiar terms with Fido. See, for example, p. 129/141: "Got back to our friends. Fido and Fido appears [*paraît*] very gay suddenly, for a week now." Or again, p. 113/124: "all the while caressing, with other hands, among other things and other words, our enclosed friend [*notre ami ci-joint*], I mean 'Fido' and Fido." (Presumably the use of singulars where one would expect plurals is because Derrida wants to blur the traditional distinction.) Or again, p. 41/47: "I had brought back, and then ordered, a whole stock of them [the postcards showing Socrates and Plato], I have two piles on the table. This morning they are two faithful dogs, Fido and Fido, two disguised children, two tired rowers." Or again, p. 178/193: "For example (I am saying this in order to reassure you: they will believe that we are two, that it's you and me, that we are legally and sexually identifiable, unless they wake up one day) in our languages, I, Fido, lack(s) [*manque*] a sex." See, finally, p. 113/125: "Basta, as Fido says, enough on this subject."

roman has to come to terms with Proust, just as anybody who wants to write a lyric in English has to come to terms with Keats.

To sum up: I am claiming that Derrida, in "Envois," has written a kind of book which nobody had ever thought of before. He has done for the history of philosophy what Proust did for his own life story: He has played all the authority figures, and all the descriptions of himself which these figures might be imagined as giving, off against each other, with the result that the very notion of "authority" loses application in reference to his work. He has achieved autonomy in the same way that Proust achieved autonomy: neither *Remembrance of Things Past* nor "Envois" fits within any conceptual scheme previously used to evaluate novels or philosophical treatises. He has avoided Heideggerian nostalgia in the same way that Proust avoided sentimental nostalgia – by incessantly recontextualizing whatever memory brings back. Both he and Proust have extended the bounds of possibility.

PART III

Cruelty and solidarity

7

The barber of Kasbeam: Nabokov on cruelty

The public-private distinction I developed throughout Part II suggests that we distinguish books which help us become autonomous from books which help us become less cruel. The first sort of book is relevant to "blind impresses," to the idiosyncratic contingencies which produce idiosyncratic fantasies. These are the fantasies which those who attempt autonomy spend their lives reworking – hoping to trace that blind impress home and so, in Nietzsche's phrase, become who they are. The second sort of book is relevant to our relations with others, to helping us notice the effects of our actions on other people. These are the books which are relevant to liberal hope, and to the question of how to reconcile private irony with such hope.

The books which help us become less cruel can be roughly divided into (1) books which help us see the effects of social practices and institutions on others and (2) those which help us see the effects of our private idiosyncrasies on others. The first sort of book is typified by books about, for example, slavery, poverty, and prejudice. These include *The Condition of the Working Class in England* and the reports of muckraking journalists and government commissions, but also novels like *Uncle Tom's Cabin, Les Misérables, Sister Carrie, The Well of Loneliness,* and *Black Boy.* Such books help us see how social practices which we have taken for granted have made us cruel.

The second sort of book – the sort I shall discuss in this chapter and the next – is about the ways in which particular sorts of people are cruel to other particular sorts of people. Sometimes works on psychology serve this function, but the most useful books of this sort are works of fiction which exhibit the blindness of a certain kind of person to the pain of another kind of person. By identification with Mr. Causaubon in *Middlemarch* or with Mrs. Jellyby in *Bleak House,* for example, we may come to notice what we ourselves have been doing. In particular, such books show how our attempts at autonomy, our private obsessions with the achievement of a certain sort of perfection, may make us oblivious to the pain and humiliation we are causing. They are the books which dramatize the conflict between duties to self and duties to others.

Books relevant to the avoidance of either social or individual cruelty are often contrasted – as books with a "moral message" – with books whose aims are, instead, "aesthetic." Those who draw this moral-aesthetic contrast and give priority to the moral usually distinguish between an essential human faculty – conscience – and an optional extra faculty, "aesthetic taste." Those who draw the same contrast to the advantage of "the aesthetic" often presuppose a distinction of the same sort. But for the latter the center of the self is assumed to be the ironist's desire for autonomy, for a kind of perfection which has nothing to do with his relations to other people. This Nietzschean attitude exalts the figure of the "artist," just as the former attitude exalts those who "live for others." It assumes that the point of human society is not the general happiness but the provision of an opportunity for the especially gifted – those fitted to become autonomous – to achieve their goal.

In the view of selfhood offered in Chapter 2, we treat both "conscience" and "taste" as bundles of idiosyncratic beliefs and desires rather than as "faculties" which have determinate objects. So we will have little use for the moral-aesthetic contrast.[1] As traditionally employed, by both "moralists" and "aesthetes," that distinction merely blurs the distinction I am trying to draw between relevance to autonomy and relevance to cruelty. The traditional picture of the self as divided into the cognitive quest for true belief, the moral quest for right action, and the aesthetic quest for beauty (or for the "adequate expression of feeling") leaves little room either for irony or for the pursuit of autonomy.[2]

If we abandon this traditional picture, we shall stop asking questions like "Does this book aim at truth or at beauty? At promoting right conduct or at pleasure?" and instead ask, "What purposes does this book serve?" Our first, broadest, classification of purpose will be built around two distinctions. The first is that between the range of purposes present-

---

1 In particular, we shall not assume that the artist must be the enemy of conventional morality. Nietzsche was unable to free himself entirely from the Kantian association of "art" and the "aesthetic," and this helped to blind him to the possibility of *liberal* ironism – a blindness sometimes shared by Bernard Shaw.

2 This Kantian-Weberian picture of three autonomous spheres is central to Habermas's work – particularly *The Theory of Communicative Action*, trans. Thomas McCarthy (Boston: Beacon Press, 1987), and *The Philosophical Discourse of Modernity*. I think Habermas is right to emphasize the way in which the separateness and autonomy of three "expert cultures" – roughly, science, jurisprudence, and literary criticism – have served the purposes of liberal society (e.g., in protecting it against would-be Lysenkos and Zhadanovs). But I think that attention to this service leads him to take an oversimple view of the relation between literature and morality – both social morality and individual morality. Habermas's classification leads him to take literature as a matter of "adequacy of the expression of feeling" and literary criticism as a matter of "judgments of taste." These notions simply do not do justice to the role which novels, in particular, have come to play in the reform of social institutions, in the moral education of the young, and in forming the self-image of the intellectual.

ly statable within some familiar, widely used, final vocabulary and the purpose of working out a *new* final vocabulary. Applying this distinction divides books up into those whose success can be judged on the basis of familiar criteria and those which cannot. The latter class contains only a tiny fraction of all books, but it also contains the most important ones – those which make the greatest differences in the long run.

The second distinction divides this latter class into those books aimed at working out a new *private* final vocabulary and those aimed at working out a new *public* final vocabulary. The former is a vocabulary deployed to answer questions like "What shall I be? "What can I become?" "What have I been?" The latter is a vocabulary deployed to answer the question "What sorts of things about what sorts of people do I need to notice?" The sort of person whom I called the "liberal ironist" in Chapter 4 needs both such vocabularies. For a few such people – Christians (and others) for whom the search for private perfection coincides with the project of living for others – the two sorts of questions come together. For most such, they do not.

The closest a liberal ironist can come to reconstructing the standard moral-aesthetic distinction, as it applies to books, is to separate books which supply novel stimuli to action (including *all* the sorts of books mentioned so far) from those which simply offer relaxation. The former suggest (sometimes straightforwardly and sometimes by insinuation) that one must change one's life (in some major or minor respect). The latter do not raise this question; they take one into a world without challenges.[3] One of the unfortunate consequences of the popularity of the moral-aesthetic distinction is a confusion of the quest for autonomy

---

3   This line between the stimulating and the relaxing, obviously, separates different books for different people. Different people lead different lives, feel challenged by different situations, and require holidays from different projects. So any attempt to go through our libraries, reshelving books with this distinction in mind, is going to be relative to our special interests. Still, it is clear that this attempt usually will *not* put Fanon's *Wretched of the Earth* and Wordsworth's *Prelude* on different shelves, nor Freud's *Introductory Lectures on Psychoanalysis* and *Middlemarch*, nor *The Education of Henry Adams* and *King Lear*, nor *A Genealogy of Morals* and the New Testament, nor Heidegger's *Letter on Humanism* and the poems of Baudelaire. So this distinction between the stimulating and the relaxing does not parallel the traditional lines between the cognitive and the noncognitive, the moral and the aesthetic, or the "literary" and the nonliterary. Nor does it conform to any standard distinctions of form or genre.

This distinction will nevertheless, for most people, separate *all* the books just mentioned from Beerbohm's *Zuleika Dobson*, Agatha Christie's *Murder on the Orient Express*, Eliot's *Old Possum's Book of Practical Cats*, Runciman's *History of the Crusades*, Tennyson's *Idylls of the King*, Saint-Simon's *Memoirs*, Ian Fleming's *Thunderball*, Macauley's *Essays*, Wodehouse's *Carry on, Jeeves!*, Harlequin romances, Sir Thomas Browne's *Urn-Burial*, and works of uncomplicated pornography. Such books gear in with their readers' fantasies without suggesting that there might be something wrong with those fantasies, or with the person who has them.

with a need for relaxation and for pleasure. This confusion is easy for those who are not ironists, and who do not understand what it is like to be an ironist – people who have never had any doubts about the final vocabulary they employ. These people – the metaphysicians – assume that books which do not supply means to the ends typically formulated in that vocabulary must be, if not immoral or useless, suitable only for private projects. Yet the only private project they can envisage is the pursuit of pleasure. They assume that a book which does supply such pleasure cannot be a serious work of philosophy, and cannot carry a "moral message." The only connection they can see between works of fiction and morality is an "inspirational" one – such works remind one of one's duty and encourage its fulfillment. This lack of understanding of irony is one reason why it is hard to convince liberal metaphysicians that some writers who give pleasure to the small group of readers who catch their allusions, and who have no relevance to liberal hope – for instance, Nietzsche and Derrida – might, nevertheless, be towering figures, capable of changing the direction of philosophical thought. It is also hard to convince liberal metaphysicians of the value of books which help us avoid cruelty, not by warning us against social injustice but by warning us against the tendencies to cruelty inherent in searches for autonomy.

In this chapter and the next I shall discuss books of this latter kind. Vladimir Nabokov and George Orwell had quite different gifts, and their self-images were quite different. But, I shall argue, their accomplishment was pretty much the same. Both of them warn the liberal ironist intellectual against temptations to be cruel. Both of them dramatize the tension between private irony and liberal hope.

In the following passage, Nabokov helped blur the distinctions which I want to draw:

> . . . *Lolita* has no moral in tow. For me a work of fiction exists only in so far as it affords me what I shall bluntly call aesthetic bliss, that is a sense of being somehow, somewhere, connected with other states of being where art (curiosity, tenderness, kindness, ecstasy) is the norm. There are not many such books. All the rest is either topical trash or what some call the Literature of Ideas, which very often is topical trash coming in huge blocks of plaster that are carefully transmitted from age to age until somebody comes along with a hammer and takes a good crack at Balzac, at Gorki, at Mann.[4]

4 Nabokov, "On a book entitled *Lolita*," in *Lolita* (Harmondsworth: Penguin, 1980), p. 313. Where it is obvious that citations are from this book, future references will be by parenthetical page number.

Orwell blurred the same distinctions when, in one of his very rare descents into rant, "The Frontiers of Art and Propaganda," he wrote exactly the sort of thing Nabokov loathed:

You cannot take a purely aesthetic interest in a disease you are dying from; you cannot feel dispassionately about a man who is about to cut your throat. In a world in which Fascism and Socialism were fighting one another, any thinking person had to take sides. . . . This period of ten years or so in which literature, even poetry, was mixed up with pamphleteering, did a great service to literary criticism, because it destroyed the illusion of pure aestheticism. . . . It debunked art for art's sake.[5]

This passage runs together two bad questions which, Nabokov rightly thought, had nothing to do with each other. The first is the question of when to take time off from private projects to resist public dangers. This question is pointless, since nobody will ever have a good *general* answer to it – although, as it happens, Orwell and Nabokov were able to agree on a particular case: Both tried vainly to enlist in the armies that were being formed to throw at the Nazis. The second question is: "Is art for the sake of art?" This is an equally bad question. It is ambiguous between "Is aesthetic bliss an intrinsic good?" and "Is aesthetic bliss the proper aim of the writer?" If the question is taken in the first sense, the answer is obviously and trivially yes. But even if we take the question in its less trivial second sense, we have to reject it. There is no such thing as "the writer," and no reason to believe that everybody who writes a book should have the same aims or be measured by the same standards.

If we firmly reject questions about the "aim of the writer" or the "nature of literature," as well as the idea that literary criticism requires taking such gawky topics seriously, we can reconcile Orwell and Nabokov in the same way I should like to reconcile Dewey and Heidegger. The pursuit of private perfection is a perfectly reasonable aim for some writers – writers like Plato, Heidegger, Proust, and Nabokov, who share certain talents. Serving human liberty is a perfectly reasonable aim for other writers – people like Dickens, Mill, Dewey, Orwell, Habermas, and Rawls, who share others. There is no point in trying to grade these different pursuits on a single scale by setting up factitious kinds called "literature" or "art" or "writing"; nor is there any point in trying to synthesize them. There is nothing called "the aim of writing" any more than there is something called "the aim of theorizing." Both Orwell and Nabokov, unfortunately, got enmeshed in attempts to excommunicate

5   George Orwell, *The Collected Essays, Journalism and Letters of George Orwell* (Harmondsworth: Penguin, 1968), vol. 2, p. 152.

people with talents and interests different from their own. This has obscured a lot of similarities between the two men, resemblances which should not be obscured by philosophical quarrels conducted in terms of factitious and shopworn oppositions like "art versus morality" or "style versus substance."

The main similarity on which I shall insist in this chapter and the next is that the books of both Nabokov and Orwell differ from those of the writers I discussed in Part II – Proust, Nietzsche, Heidegger, and Derrida – in that cruelty, rather than self-creation, is their central topic. Both Nabokov and Orwell were political liberals, in a broad sense of the term which can be stretched to cover Proust and Derrida (though not to cover either Nietzsche or Heidegger). They shared roughly the same political credo, and the same reactions to the same political events. More important, however, they both met Judith Shklar's criterion of a liberal: somebody who believes that cruelty is the worst thing we do.[6] Nabokov wrote about cruelty from the inside, helping us see the way in which the private pursuit of aesthetic bliss produces cruelty. Orwell, for the most part, wrote about cruelty from the outside, from the point of view of the victims, thereby producing what Nabokov called "topical trash" – the kind of book which helps reduce future suffering and serves human liberty. But I shall argue in Chapter 8 that at the end of his last book, in his portrait of O'Brien, Orwell does the same as Nabokov: He helps us get *inside* cruelty, and thereby helps articulate the dimly felt connection between art and torture.

In the remainder of this chapter I shall offer a reading of Nabokov which connects three of his traits: his aestheticism, his concern with cruelty, and his belief in immortality. "We believe ourselves to be mortal," Nabokov writes, "just as a madman believes himself to be God."[7]

To see what Nabokov's aestheticism looked like when he was dealing with an author whom he took seriously, consider his lecture on Dickens's *Bleak House*. At one point he quotes at length from the chapter where Dickens describes the death of the boy Jo. This is the chapter whose coda is the famous paragraph beginning "Dead, your Majesty! Dead, my lords and gentlemen!" and ending "And dying around us every day." That is a call to public action if anything in Dickens is. But Nabokov tells us that the chapter is "a lesson in style, not in participative emotion."[8]

6 See Judith Shklar, *Ordinary Vices*, pp. 43–44, and Chapter 1, passim.
7 "Comme un fou se croit Dieu, nous nous croyons mortels" is the epigraph to *Invitation to a Beheading*. Nabokov attributes the sentence to "the melancholy, extravagant, wise, witty, magical, and altogether delightful Pierre Delalande, whom I invented."
8 Vladimir Nabokov, *Lectures on Literature*, ed. Fredson Bowers (New York: Harcourt

Notice that if Nabokov had said "as well as" instead of "not," nobody would have disagreed. By saying "not" he maintains his stance as someone who is concerned with nothing but "aesthetic bliss," someone who thinks that "the study of the sociological or political impact of literature has to be devised mainly for those who are by temperament or education immune to the aesthetic vibrancy of authentic literature, for those who do not experience the telltale tingle between the shoulder blades" (*LL*, p. 64). Nabokov has to pretend, implausibly, that Dickens was not, or at least should not have been, interested in the fact that his novels were a more powerful impetus to social reform than the collected works of all the British social theorists of his day.

Why does Nabokov insist that there is some incompatibility, some antithetical relation, between Housmanian tingles and the kind of participative emotion which moved liberal statesmen, such as his own father, to agitate for the repeal of unjust laws? Why doesn't he just say that these are two distinct, noncompetitive, goods? Nabokov is quite right when he says, "That little shiver behind is quite certainly the highest form of emotion that humanity has attained when evolving pure art and pure science" (*LL*, p. 64). This dictum simply spells out the relevant sense of the term "pure." But it seems quite compatible with saying that the ability to shudder with shame and indignation at the unnecessary death of a child – a child with whom we have no connection of family, tribe, or class – is the highest form of emotion that humanity has attained while evolving modern social and political institutions.

Nabokov does not try to defend his assumption that social reform does not have the same claim on our attention as "pure art and pure science." He gives no reasons for doubting that people as gifted as Dickens have sometimes been able to do quite different things in the same book. It would have been much easier to admit that *Bleak House* aroused participative emotions which helped change the laws of England, and *also* made Dickens immortal by having been written so as to keep right on producing tingles between the shoulder blades long after the particular horrors of Dickens's century had been replaced by new ones. Yet Nabokov insists over and over again that the latter accomplishment – the effect produced by style as opposed to that produced by participative emotion – is *all* that matters.[9] He never makes clear what scale

Brace Jovanovich, 1980), p. 94. Henceforth this book will be cited parenthetically as "*LL*."

9 He never quite brings himself to say that artists should not pay attention to social evils or try to change them. But he is churlish about any given attempt to do so, and often on wildly irrelevant grounds. He says, inaccurately and pointlessly, that "the link of these poor children in *Bleak House* is not so much with social circumstances of the 1850s as with earlier times and mirrors of time." With equal irrelevance, he dismisses the chap-

of importance he is using, nor why we should insist on a *single* scale. It is hardly *evident* that "pure art and pure science" matter more than absence of suffering, nor even that there is a point in asking which matters more – as if we could somehow rise above both and adjudicate their claims from a neutral standpoint.

I share Nabokov's suspicion of general ideas when it comes to philosophers' attempts to squeeze our moral sentiments into rules for deciding moral dilemmas. But I take the lesson of our failure to find such rules to be that we should stop talking in a quasi-metaphysical style about the "task of the writer" or "what ultimately matters," or the "highest emotion"; stop working at the level of abstraction populated by such pallid ghosts as "human life," "art," and "morality"; and stay in a middle range. We should stick to questions about what works for particular purposes. So as a first stage in reconciling Orwell and Nabokov I would urge that Orwell shares some important purposes with Dickens (producing shudders of indignation, arousing revulsion and shame), and Nabokov shares others (producing tingles, aesthetic bliss).

But Nabokov does not want to be reconciled. He wants Dickens and himself to count as members of an elect from which Orwell – and other objects of his contempt, such as Balzac, Stendhal, Zola, Gorki, Mann, Faulkner, and Malraux – are forever excluded. We get an important clue to his motives from a passage in which he explains why he reads Dickens as he does:

As is quite clear, the enchanter interests me more than the yarn spinner or the teacher. In the case of Dickens, this attitude seems to me to be the only way of keeping Dickens alive, above the reformer, above the penny novelette, above the sentimental trash, above the theatrical nonsense. There he shines forever on the heights of which we know the exact elevation, the outlines and the formation, and the mountain trails to get there through the fog. It is in his imagery that he is great. (*LL,* p. 65)

The fog in question is the one Dickens has described in the opening chapter of *Bleak House.* As Nabokov says, Dickens uses the London fog to revivify a standard trope: the legal miasma which rises from proceedings in Chancery. Nabokov wants us to treat Dickens's attacks on the evil of the Chancery system – and more generally his portrayal of conflicts between what Nabokov, putting the words in shudder quotes, calls "good" and "evil" –as merely the "skeleton" of *Bleak House.* He congrat-

ters about Sir Leicester and Lady Dedlock, considered as an "indictment of the aristocracy," as of "no interest or importance whatsoever since our author's knowledge and notions of that set are extremely meager and crude" (*LL,* pp. 64–65).

ılates Dickens on being "too much of an artist" to make this skeleton
'obstrusive or obvious." Writers without Dickens's ability, the people
vho write "topical trash," do not know who to put flesh on the "moral"
ıkeleton of their work. So, to mix the two metaphors, heaps of such
ıiled-up skeletons – the novels of Orwell and Mann, for example – form
:he fogbound, boggy foothills of literature. For lack of precise imagery,
writers who can give lessons in participative emotion but not in style fail
:o achieve immortality.

Two things should be noticed about the passage I have just quoted.
The first is that Nabokov is writing about Dickens not for the sake of the
students in his class, nor for the sake of the educated public, but *solely* for
Dickens's sake. He wants to do a favor for one of his few peers. He wants
him to have the immortality he deserves. When he says, for example,
that Edmund Wilson's treatment of Dickens in *The Wound and the Bow* is
"brilliant" but that the "sociological side" of Dickens is "neither interesting nor important," he is saying that literary criticism of the sort which
Wilson did brilliantly creates the same kind of particularly thick fog as
was created by particularly brilliant members of the Chancery Bar.[10] By
pointing out the mountain peak above the fog, and by tracing the trails
that reach it, he is rescuing Dickens from people like Wilson, rescuing
him from the creeping miasma of historical time and mortal chance.

The second thing to notice is that Nabokov's concern with Dickens's
immortality was a corollary of his own intense, lifelong preoccupation
with the question of whether he might survive death, and thereby meet
his parents in another world. Such survival, and such meetings, suddenly
appear, in the last lines of *Invitation to a Beheading*, as the point of that
novel. They are also the topic of a canto of John Shade's poem "Pale
Fire," and of the magnificent closing sentences of *Lolita:*

And do not pity C.Q. [Clare Quilty]. One had to choose between him and H.H.
[Humbert Humbert], and one wanted H.H. to exist at least a couple of months
longer, so as to have him make you live in the minds of later generations. I am
thinking of aurochs and angels, the secret of durable pigments, prophetic son-

10 Nabokov must, when he read *The Wound and the Bow*, have realized how easily
Wilson's general strategy – tracing a writer's obsession, and the shape of his career, to
some early injury – could be applied to his, Nabokov's, own case. He must have been
infuriated by the realization that this application would have already occurred to
Wilson. I suspect that he was also annoyed by Wilson's quasi-Freudian interpretation
of Housman. Nabokov's talk of tingles was certainly influenced by Housman's *Name
and Nature of Poetry* (the best-known manifesto in English of what Nelson Goodman
calls "the Tingle-Immersion" theory of aesthetic experience). He liked Housman's
poetry when he was an undergraduate at Trinity, although later he unkindly refers to
A Shropshire Lad as a "little volume of verse about young males and death."

nets, the refuge of art. And this is the only immortality you and I may share, my Lolita (p. 307).

In this latter passage, as in many others, Nabokov is talking about immortality in the "literary" sense – the sense in which one is immortal if one's books will be read forever. But elsewhere, especially in his autobiography, he talks about immortality in the ordinary theological and metaphysical sense – the chance of somehow surviving death, and of thus being able to meet dead loved ones in a world beyond time.[11] He makes no bones about his own fear of death (*SM*, p. 80):

Over and over again, my mind has made colossal efforts to distinguish the faintest of personal glimmers in the impersonal darkness on both sides of my life. That this darkness is caused merely by the walls of time separating me and my bruised fists from the free world of timelessness is a belief I gladly share with the most gaudily painted savage. (*SM*, p. 14)

Over and over again, Nabokov tried to tie this highly unfashionable concern for metaphysical immortality together with the more respectable notion of literary immortality. He wanted to see some connection between creating tingles, creating aesthetic bliss, being an artist in the sense in which he and Joyce and Dickens were artists and Orwell and Mann were not, and freeing oneself from time, entering another state of being. He is sure that there is a connection between the immortality of the work and of the person who creates the work – between aesthetics and metaphysics, to put it crudely. But, unsurprisingly, he is never able to say what it is.

The best example of this gallant, splendid, and foredoomed effort is one of Nabokov's few attempts to work in the uncongenial medium of general ideas. This is his essay called "The Art of Literature and Common Sense," in which he offers the same generalized protest against general ideas which we find in Heidegger. Heidegger and Nabokov agree that common sense is a self-deceptive apologia for thoughtlessness and vulgarity. They offer the same defense of unique and idiosyncratic irony. They both reject the Platonist and democratic claim that one should only have beliefs which can be defended on the basis of widely shared premises. The theme of Nabokov's essay is what he calls the "supremacy of the detail over the general" (*LL*, p. 373). His thesis is: "This capacity to wonder at trifles – no matter the imminent peril –

11   Vladimir Nabokov, *Speak, Memory: An Autobiography Revisited* (New York, Pyramid, 1968), pp. 14, 37, 57, 87, 103. Henceforth this book will be cited parenthetically as "*SM*."

these asides of the spirit, these footnotes in the volume of life are the highest forms of consciousness, and it is in this childishly speculative state of mind, so different from commonsense and its logic, that we know the world to be good" (*LL,* p. 374).

Here we are not told merely, and tautologously, that "pure art and pure science" culminate in such tingling trifles. We are told that these tingles are "the highest forms of consciousness." That claim is ambiguous between a moral and a metaphysical interpretation. It can mean that tingles are what is most worth striving for, or it can mean the sort of thing Plato meant, that this form of consciousness is higher in that it gets us in touch with the nontemporal, in that it gets us out of the flux and into a realm beyond time and chance. If one took the claim only in its moral sense, then one could plausibly reply that this was certainly what it behooved people like Nabokov to strive for, but that other people with other gifts – people whose brains are not wired up to produce tingles, but who are, for example, good at producing shudders of moral indignation – might reasonably strive for their own form of perfection. But Nabokov wanted to absolutize the moral claim by backing it up with the metaphysical claim. He wanted to say that idiosyncratic imagery, of the sort he was good at, rather than the kind of generalizing ideas which Plato was good at, is what opens the gates of immortality. Art, rather than mathematics, breaks through the walls of time into a world beyond contingency.

The trouble with the essay is, once again, that Nabokov runs together literary with personal immortality. If only the former is at stake, then, indeed, Plato was wrong and Nabokov, Heidegger, and Derrida are right.[12] If you want to be remembered by future generations, go in for

12   What we have learned since Plato is that general ideas are tools for practical purposes, purposes which are forgotten as time goes by, but that particular images survive. Nowadays we can do better in the way of moral ideals, social arrangements, and human beings than Homer imagined. As Nabokov puts it, "In the imaginary battle of [*homo*] *americus* versus [*homo*] *homericus,* the first wins humanity's prize." But Homer survives because his images survive. Boys who adopt Achilles' ethic ("always outdo the others") are just boring bullies, but certain Homeric epithets still make their quieter classmates tingle. Nobody knows, or much cares, whether Shakespeare wanted to get across a sociopolitical view in the Roman plays; but John Shade speaks for all us anti-Platonists when, in response to Kinbote's suggestion that he "appreciates particularly the purple passages" in *Hamlet,* he replies: "Yes, my dear Charles, I roll upon them as a grateful mongrel on a spot of turf fouled by a Great Dane." The question of retirement to a Sabine farm has gone stale, but we bow toward Horace whenever we describe a passage as purple. Plato himself, though generally wrong about general ideas, survives as the first white magus. He is the enchanter who spun the first strands of that web of metaphor which Derrida calls the West's "white mythology." His own special fire still smolders and his particular sun still blazes, long after the mathematics he admired has become a tool in the hands of bomb builders,

poetry rather than for mathematics. If you want your books to be read rather than respectfully shrouded in tooled leather, you should try to produce tingles rather than truth. What we call common sense – the body of widely accepted truths – is, just as Heidegger and Nabokov thought, a collection of dead metaphors. Truths are the skeletons which remain after the capacity to arouse the senses – to cause tingles – has been rubbed off by familiarity and long usage. After the scales are rubbed off a butterfly's wing, you have transparency, but not beauty – formal structure without sensuous content. Once the freshness wears off the metaphor, you have plain, literal, transparent language – the sort of language which is ascribed not to any particular person but to "common sense" or "reason" or "intuition," ideas so clear and distinct you can look right through them. So if, like Euclid's or Newton's or J. S. Mill's, your metaphors are socially useful and become literalized, you will be honored in the abstract and forgotten in the particular. You will have become a name but ceased to be a person. But if, like Catullus, Baudelaire, Derrida, and Nabokov, your works (only, or also) produce tingles, you have a chance of surviving as more than a name. You might be, like Landor and Donne, one of the people whom some future Yeats will hope to dine with, at journey's end.

However, although all this is quite true, it has no bearing on the suggestion that literary immortality is connected with personal immortality – the claim that you will actually *be* out there, beyond the walls of time, waiting for dinner guests. As Kant pointed out, and as Nabokov ruefully admitted, nothing *could* lend plausibility to that claim. Waiting, like everything else one can imagine doing, takes time.[13] But even if we dismiss the metaphysical claim, we still need to take seriously a further claim Nabokov makes – that it is in "this childishly speculative state of mind" that "we know the world to be good."

Nabokov thinks that "goodness" is something irrationally concrete, something to be captured by imagination rather than intellect. He inverts Plato's divided line so that *eikasia,* rather than *nous,* becomes the faculty of moral knowledge.[14] He says:

and long after the moral intuitions he hoped to refine and purify have been exposed as the inconsistent proverbs of a rather primitive culture. In respect to what Whitehead called "objective immortality," the great figures of the past are indeed, as Nabokov says of Dickens, "great in their imagery."

13   Nabokov was as resigned as Kant to the fact that Swedenborgesque speculation will never get anywhere: "I have journeyed back in time – with thought hopelessly taper-ing off as I went – to remote regions where I groped for some secret outlet only to discover that the prison of time is spherical and without exits" (*SM,* p. 14).

14   Nabokov may have been influenced by Bergson's attempt at an inverted Platonism, and in particular by *The Two Sources of Morality and Religion.* See Nabokov, *Strong*

From the commonsensical point of view the "goodness," say, of some food is just as abstract as its "badness," both being qualities that cannot be perceived by the sane judgment as tangible and complete objects. But when we perform that necessary mental twist which is like learning to swim or to make a ball break, we realize that "goodness" is something round and creamy, and beautifully flushed, something in a clean apron with warm bare arms that have nursed and comforted us. (*LL*, p. 375)

In the same essay he brings together this idea of the good as something "real and concrete" with his sense of solidarity with a "few thousand" others who share his gifts:

. . . the irrational belief in the goodness of man . . . becomes something much more than the wobbly basis of idealistic philosophies. It becomes a solid and iridescent truth. This means that goodness becomes a central and tangible part of one's world, which world at first sight seems hard to identify with the modern one of newspaper editors and other bright pessimists, who will tell you that it is, mildly speaking, illogical to applaud the supremacy of good at a time when something called the police state, or communism, is trying to turn the globe into five million square miles of terror, stupidity, and barbed wire. . . . But within the emphatically and unshakably illogical world which I am advertising as a home for the spirit, war gods are unreal not because they are conveniently remote in physical space from the reality of a reading lamp and the solidity of a fountain pen, but because I cannot imagine (and that is saying a good deal) such circumstances as might impinge upon the lovely and lovable world which quietly persists, whereas I can very well imagine that my fellow dreamers, thousands of whom roam the earth, keep to these same irrational and divine standards during the darkest and most dazzling hours of physical danger, pain, dust, death. (*LL*, p. 373)

I interpret these two passages as making an important psychological point: that the only thing which can let a human being combine altruism and joy, the only thing that makes either heroic action or splendid speech possible, is some very specific chain of associations with some highly idiosyncratic memories.[15] Freud made the same point, and Freud

Opinions (London: Weidenfeld & Nicolson, 1974), pp. 42, 290. Henceforth cited parenthetically as "*SO*."

15 In this sense, though in no other, Nabokov is right in saying that "everybody thinks in images and not in words" (*SO*, p. 14). I would argue that if you can't use language, you can't be conscious of inner images any more than of outer objects, but this Sellarsian "psychological nominalist" thesis is compatible with agreeing that what individuates people, gives them their special flavors and their distinctive neuroses, are not different propositional attitudes but different associations of the words in their final vocabularies (including the word "good," which occurs in almost everybody's) with particular situations. In the case of special sorts of people like Nabokov, who have specially wired brains, this means association with extremely vivid and detailed images of those

was the one person Nabokov resented in the same obsessive and intense way that Heidegger resented Nietzsche. In both cases, it was resentment of the precursor who may already have written all one's best lines. This psychological thesis binds Hume, Freud, and Nabokov and distinguishes them from Plato and Kant. But it is neither a metaphysical claim about the "nature" of "goodness" nor an epistemological claim about our "knowledge" of "goodness." Being impelled or inspired by an image is not the same as knowing a world. We do not need to postulate a world beyond time which is the home of such images in order to account for their occurrence, or for their effects on conduct.

Yet only if he could somehow have squeezed some metaphysics out of his two soundly anti-Platonic claims – the one about the nature of literary immortality and the other about the nature of moral motivation – would Nabokov have been able to hook up the utilization of his own gifts with the nature of things. Only then could he see his special gifts as putting him in an epistemologically privileged position, in a position to be aware of the secret that, as gaudily painted savages believe and as Cincinnatus C. eventually realized, time and causality are merely a vulgar hoax. Only if he can make such a hookup will he be able to defend his claim that Dickens's interest in social reform was merely a great artist's foible, and his suggestion that topical writers like Orwell deserve no thanks for their services to human liberty.

The collection of general ideas which Nabokov assembled in the hope of convincing himself that time and causality *were* hoaxes is an odd, inconsistent mixture of Platonic atemporalism and anti-Platonic sensualism. It is an attempt to combine the comforts of old-fashioned metaphysics with the up-to-date antimetaphysical polemic common to Bergson and Heidegger. Like the systems of general ideas which ironist theorists construct in order to attack the very idea of a general idea, it is what Stanley Fish calls a "self-consuming artifact." Still, such fragile and unbalanced devices, with their artful combination of dogma and irony, have the same iridescence as John Shade's poem, "Pale Fire." Like that poem, Nabokov's system is the shadow of the waxwing, just before it smashes itself against the walls of time.

Why did Nabokov want such a device? Why did he stick his neck out in this way? I think there were two reasons, neither of which had anything to do with fear of death. The first, and most important, was an oversize sense

situations. But Nabokov, of course, goes overboard when he claims that people without his special eidetic faculty lead simple and vulgar lives. There are lots of ways for a mind to be rich and interesting which do not involve imagery (e.g., music – to which Nabokov, like Yeats and unlike his own father and his own son, was almost completely insensitive).

of pity. His eccentrically large capacity for joy, his idiosyncratic ability to experience bliss so great as to seem incommensurable with the existence of suffering and cruelty, made him unable to tolerate the reality of suffering. Nabokov's capacity to pity others was as great as Proust's capacity to pity himself – a capacity which Proust was, amazingly, able to harness to his attempt at self-creation. Bliss began early for Nabokov. He had no occasion for self-pity and no need for self-creation. The difference between Proust's novel and Nabokov's novels is the difference between a bildungsroman and a crescendo of ever more fervent pledges of the same childhood faith. Nabokov seems never to have suffered a loss for which he blamed himself, never to have despised, distrusted, or doubted himself. He did not *need* to struggle for autonomy, to forge a conscience in the smithy of his soul, to seek a self-made final vocabulary. He was a hero both to his parents and to himself – a very lucky man. He would have been merely a self-satisfied bore if it were not that his brain happened to be wired up so as to make him able continually to surprise and delight himself by arranging words into iridescent patterns.

But the other side of this capacity for bliss was an inability to put up with the thought of intense pain. The intensity of his pity brought him to the novel which has aroused most protest among his admirers: *Bend Sinister*. In this novel, the eight-year-old son of Adam Krug is tortured to death by madmen because his folder has been misfiled by the inexperienced bureaucrats of a revolutionary government. Nabokov does not attempt to portray Krug's pain. More than that, he refuses to countenance the reality of a pain that great. So, as in *Invitation to a Beheading*, he translates the hero to another "realm of being." In the earlier novel, Cincinnatus rises as soon as his head has been chopped off, watches the scaffold and spectators dissolve, and then "makes his way in that direction where, to judge by the voices, stood beings akin to him." In *Bend Sinister*, Nabokov saves Krug from the realization of what has happened by what he calls the "intervention of an anthropomorphic deity impersonated by me."[16] Nabokov says he "felt a pang of pity for Adam and slid towards him along an inclined beam of pale light – causing instantaneous madness, but at least saving him from the senseless agony of his logical fate" (pp. 193–194). Krug's author steps through "a rent in his [Krug's] world leading to another world of tenderness, brightness, and beauty" (p. 8). Nabokov's toying with general ideas about immortality, with the idea that there was a rent in his and our worlds like that in Krug's, was a further expression of the same pity which saved Cincinnatus and Krug.

16   Vladimir Nabokov, *Bend Sinister* (Harmondsworth: Penguin, 1974), p. 11. Henceforth cited parenthetically by page number.

But there is a second reason which needs to be taken into account. This is that Nabokov seems never to have allowed himself social hope. He was the son of a famous liberal statesman who was assassinated when his son was twenty-two. His father's circle – which included, for example, H. G. Wells, whom Nabokov met at his father's table – had no time for metaphysics because their hopes were centered on future generations. They exemplified the substitution of hope for future generations for hope of personal immortality which I discussed in Chapter 4. It was just this former hope that Nabokov seems to have had no trace of. Perhaps he had it once, and abandoned it as a result of his father's murder. Perhaps he never had it, having recognized early on that he and his father had antithetical, if equally great, gifts, and that he would betray himself by attempting even the slightest imitation of someone he loved so fiercely. Whatever the reason, he always repudiated any interest in political movements. In *The Gift,* Fyodor walks down the streets of Berlin in the 1920s and notices that "three kinds of flags were sticking out of the house windows: black-yellow-red, black-white-red, and plain red: each now meant something, and funniest of all, this something was able to excite pride or hatred in someone." Noticing the flags induces a meditation on Soviet Russia, which ends when Fyodor thinks:

Oh, let everything pass and be forgotten – and again in two hundred years' time an ambitious failure will vent his frustration on the simpletons dreaming of a good life (that is if there does not come *my* kingdom, where everyone keeps to himself and there is no equality and no authorities – but if you don't want it, I don't insist and don't care).[17]

Nabokov had no idea – who does? – about how to bring about a state with no equality and no authorities. But he also gave up on the modern liberal idea of working for a future in which cruelty will no longer be institutionalized. In this respect, he was a throwback to antiquity, to a time when such social hope was so obviously unrealistic as to be of little interest to intellectuals. His otherworldly metaphysics is what one might imagine being written by a contemporary of Plato's, writing in partial imitation of, and partial reaction against, the *Phaedo* – a contemporary who did not share Plato's need for a world in which he could not feel shame, but did need a world in which he would not have to feel pity.

If, however, Nabokov's career as a novelist had climaxed with the creation of Fydor Godunov-Cherdyntsev, Cincinnatus C., and Adam Krug, we would not read him as often as we do. The characters I just

17 Vladimir Nabokov, *The Gift* (Harmondsworth: Penguin, 1963), p. 370.

mentioned are known because they were created by the author of two others – Humbert Humbert and Charles Kinbote. These two are the central figures of Nabokov's books about cruelty – not the "beastly farce" common to Lenin, Hitler, Gradus, and Paduk, but the *special* sort of cruelty of which those capable of bliss are also capable. These books are reflections on the possibility that there can be sensitive killers, cruel aesthetes, pitiless poets – masters of imagery who are content to turn the lives of other human beings into images on a screen, while simply not noticing that these other people are suffering. Nabokov's uneasiness at the unstable philosophical compromise which he had worked out, and what must have been at least occasional doubts about his refusal to think in terms of human solidarity, led him to consider the possibility that he was mistaken. Like the honest man he was, Nabokov wrote his best books to explore the possibility that his harshest critics might, after all, be right.

What his critics were suggesting was that Nabokov was really Harold Skimpole. Skimpole, the charming aesthete in *Bleak House*, brings about Jo's death – an action beautifully described by Nabokov as "the false child betraying the real one" (*LL*, p. 91). Skimpole claims the privileges of the child and of the poet. He views everyone else's life as poetry, no matter how much they suffer.[18] Skimpole sees his having taken five pounds to betray Jo's whereabouts to Tulkinghorn's agent as an amusing concatenation of circumstances,[19] a pleasant little poem, the sort of thing John Shade calls "some kind of link-and-bobolink, some kind of corre-lated pattern in the game." By claiming not to grasp concepts like "mon-ey" and "responsibility," Skimpole tries to exonerate himself from living off the charity and the suffering of others.[20]

It is clear from his autobiography that the only thing which could really get Nabokov down was the fear of being, or having been, cruel. More specifically, what he dreaded was simply not having *noticed* the suffering of someone with whom one had been in contact (*SM*, pp. 86–87). It hurt Nabokov horribly to remember the pain he might unthinkingly have caused to a schoolmate, or a governess. It must have terrified him to think that he might be Skimpole after all. The intensity of his fear of cruelty seems to me to show that we should read *Pale Fire* as about two of Nabokov's own personae. On the one side there is John Shade, who combines Nabokov's private virtues with a Jarndyce-like patience for his

18   Charles Dickens, *Bleak House* (New York: Signet, 1964), pp. 445, 529.
19   See Nabokov's discussion of this passage in *LL*, p. 90.
20   This was sometimes Edmund Wilson's view of what his friend Nabokov was doing. Wilson occasionally cast himself in the part of John Jarndyce, the patient and generous patron, opposite Nabokov as the charmingly amoral Skimpole.

monstrous friend, the false child Kinbote. On the other is Kinbote himself, whose central characteristic is his inability to notice the suffering of anyone else, especially Shade's own, *but who is a much better writer than Shade himself.*

Nabokov's greatest creations are obsessives – Kinbote, Humbert Humbert, and Van Veen – who, *although they write as well as their creator at his best,* are people whom Nabokov himself loathes – loathes as much as Dickens loathed Skimpole. Humbert is, as Nabokov said, "a vain and cruel wretch who manages to appear 'touching'" (*SO,* p. 94) – manages it because he can write as well as Nabokov can. Both Kinbote and Humbert are exquisitely sensitive to everything which affects or provides expression for their own obsession, and entirely incurious about anything that affects anyone else. These characters dramatize, as it has never before been dramatized, the particular form of cruelty about which Nabokov worried most – incuriosity.

Before giving examples from the novels of this cruel incuriosity, let me offer another sort of evidence to back up the claim I have just made. Remember Nabokov's rapid parenthetical definition of the term "art" in the passage about "aesthetic bliss" cited early in this chapter. Writing what he knew would be the most discussed passage of what he knew would become his most widely read manifesto, the Afterword to *Lolita,* he identifies art with the compresence of "curiosity, tenderness, kindness, and ecstasy." Notice that "curiosity" comes first.[21]

Nabokov is, I think, trying to jam an ad hoc and implausible moral philosophy into this parenthesis, just as he is trying to jam metaphysical immortality into the phrase "other states of being," which he uses to define "aesthetic bliss." If curiosity and tenderness are the marks of the artist, if both are inseparable from ecstasy – so that where they are

---

21 The background of this definition of art is interesting. Since Nabokov seems never to have forgotten anything, it is likely that a snide remark to Wilson about "the stale Bolshevik propaganda which you imbibed in your youth" (*The Nabokov-Wilson Letters,* ed. Simon Karlinksy [New York: Harper, 1979], p. 304; December 13, 1956) was an allusion to an equally snide remark Wilson had made eight years previously. In 1948, Wilson had written Nabokov as follows: "I have never been able to understand how you manage, on the one hand, to study butterflies from the point of view of their habitat and, on the other, to pretend that it is possible to write about human beings and leave out of account all question of society and environment. I have come to the conclusion that you simply took over in your youth the *fin de siècle Art for Art's sake* slogan and have never thought it out. I shall soon be sending you a book of mine [*The Triple Thinkers*] which may help you to straighten out these problems" (ibid., p. 211; November 15, 1948). Nabokov replied immediately. After dismissing Faulkner and Malraux, two of Wilson's favorites, as "popular mediocre writers," he says, "'Art for art's sake' does not mean anything unless the term 'art' be defined. First give me your definition of it and then we can talk" (ibid., p. 214; November 21, 1948). Wilson did not take up this challenge, but Nabokov did, in the passage I have been discussing.

NABOKOV ON CRUELTY

absent no bliss is possible – then there is, after all, no distinction between the aesthetic and the moral. The dilemma of the liberal aesthete is resolved. All that is required to act well is to do what artists are good at – noticing things that most other people do not notice, being curious about what others take for granted, seeing the momentary iridescence and not just the underlying formal structure. The curious, sensitive artist will be the paradigm of morality because he is the only one who always notices everything.

This view is, once again, an inverted Platonism: Plato was right that to know the good is to do it, but he gave exactly the wrong reason. Plato thought that "knowing the good" was a matter of grasping a general idea, but actually knowing the good is just sensing what matters to other people, what their image of the good is – noticing whether they think of it as something round and creamy and flushed, or perhaps as something prism-shaped, jewel-like, and glistening. The tender, curious artist would be the one who, like Shade and unlike Skimpole or Kinbote, has time for other people's fantasies, not just his own. He would be a nonobsessed poet, but nonetheless one whose poems could produce ecstasy.[22]

But Nabokov knew quite well that ecstasy and tenderness not only are separable but tend to preclude each other – that most nonobsessed poets are, like Shade, second rate. This is the "moral" knowledge that his novels help us acquire, and to which his aestheticist rhetoric is irrelevant. He knows quite well that the pursuit of autonomy is at odds with feelings of solidarity. His parenthetical moral philosophy would be sound only if it were true that, as Humbert says, "poets never kill." But, of course, Humbert does kill – and, like Kinbote, Humbert is exactly as good a writer, exactly as much of an artist, capable of creating exactly as much iridescent ecstasy, as Nabokov himself. Nabokov would like the four characteristics which make up art to be inseparable, but he has to face up to the unpleasant fact that writers can obtain and produce ecstasy while failing to notice suffering, while being incurious about the people whose lives provide their material. He would like to see all the evil in the world – all the failures in tenderness and kindness – as produced by nonpoets, by generalizing, incurious vulgarians like Paduk and Gradus.[23] But he

---

22 Contrast Nabokov's list of the characteristics of art with what Baudelaire tells us is the norm in Cythera, namely, "order, beauty, voluptuous luxury and calm." This was also the norm of the country houses of Nabokov's childhood, islands in the middle of what Nabokov says was a "to be perfectly frank, rather appalling country" (SM, pp. 85–86). Nabokov's definition gives a new twist to the slogan "art for art's sake," and to the relation between art and morality. Baudelaire's description of the Cytheran norm does not mention any relation to other human beings, except perhaps voluptuous enjoyment. But Nabokov's does.
23 Nabokov might have included Lenin on this list. But he might not have, since he must

knows that this is not the case. [24] Nabokov would desperately like artistic gifts to be sufficient for moral virtue, but he knows that there is no connection between the contingent and selective curiosity of the autonomous artist and his father's political project – the creation of a world in which tenderness and kindness are the human norm. So he creates characters who are both ecstatic and cruel, noticing and heartless, poets who are only selectively curious, obsessives who are as sensitive as they are callous. [25] What he fears most is that one cannot have it both ways – that there is no synthesis of ecstasy and kindness.

have suspected that he and Lenin himself had more in common than either did with Paduk or Gradus. Lenin, I think, hovers in the background of Nabokov's consciousness as the terrifying O'Brien figure – the man who will rule the world because he combines Paduk's cruelty with something uncomfortably like Nabokov's brains. Nabokov's official position is that "Lenin's life differs from, say, James Joyce's as much as a handful of gravel does from a blue diamond, although both men were exiles in Switzerland and both wrote a vast number of words" (SO, pp. 118–119). But I doubt that he could really bring himself to believe that.

24   He knows not only that Humbert is wrong when he says that "poets never kill" but that it is pointless to say, with Kinbote, that "the one who kills is always his victim's inferior" (Pale Fire, p. 157). For "inferiority" means nothing here – it is one of those machine-made general ideas. If we could specify in what respect murderers were always inferior to their victims, in what respect, for example, Humbert is inferior to Quilty or O'Brien to Winston, then we might have said something useful. But all we can say is that they are morally inferior – and if that is what we mean, it would be better to say, "Thou shalt not kill," and have done with it. Nabokov's point about general ideas is that once the concrete detail is left behind, everything quickly blurs together, and the result might as well be left unsaid.

25   What makes Humbert and Kinbote such interesting people is that, although they rarely react to people in predictable ways, they are not oblivious of other people. Not only are they intensely, albeit selectively, curious, but their own minds find a "kind of twisted pattern in the game," a motif in the lives of others. The question of whether that pattern was really there is as bad a question as that of whether an artist "truly represents" human emotions. Once the artist has done his work, it is as much "there" as the pattern which conventional moral discourse finds in the same story of joy and suffering. Kinbote is not "making something up" when he reads the story of Zembla between the lines of Shade's poem, any more than he is "representing inaccurately." He is reacting to a stimulus, and thereby creating a new stimulus.

It is important to see that Kinbote cares a great deal about Shade's poem, even if for all the wrong reasons. He thinks very hard about it, even though his thought goes in utterly different directions from Shade's. This illustrates the point that a perverse, egocentric commentary – what Bloom calls a "strong reading" – is still a commentary. Just as a Heideggerian reading of Kant is still a reading of Kant, so the reaction of someone like Kinbote to the suicide of Shade's daughter is something we have to take into account. It is something we ought to be curious about, just as we ought to be curious about John Shade's reaction, which was to write "Pale Fire," or Sybil Shade's, which was to translate Donne's sermon on death and Marvell's "The Nymph on the Death of her Fawn" into French. (See Nabokov, Pale Fire [New York: Berkeley, 1968], pp. 33, 161–162.) People react to intolerable ecstasy or hopeless longing or intense pain as best they can, and once we leave the realm of action for that of writing, it is no service to anyone to ask whether a reaction was "appropriate." For appropri-

The two novels of his acme spell out this fear.[26] The remarkable thing about both novels is the sheer originality of the two central characters – Humbert and Kinbote. No one before had thought of asking what it would be like to be a Skimpole who was also a genius – one who did not simply toss the word "poetry" about but who actually *knew* what poetry was. This particular sort of genius-monster – the monster of incuriosity – is Nabokov's contribution to our knowledge of human possibilities. I

---

ateness is a matter of taking up a place within a preestablished and familiar pattern. The curiosity which Nabokov thought essential to art consists in never being content with such a pattern.

26 I shall use a footnote to say something about why I take *Lolita* and *Pale Fire* as Nabokov's acme. I am urging that we think of these two novels as revolving around the same theme as Nabokov's early novel *The Gift* – namely, around the choice between tenderness and ecstasy which those gifted with artistic talent face, the necessity that they be only *selectively* curious. Compared to these two later novels, however, *The Gift* is didactic, a set of illustrations for some general ideas. The trajectory of Nabokov's career, like that of Heidegger's, was shaped by the attempt to avoid being didactic, to avoid the use of words which had been tarnished, reduced to near transparency, by common use. Nabokov criticized his first novel, *Mary*, by saying that "the émigré characters I had collected in that display box [*Mary*] were so transparent to the eye of the era that one could easily make out the labels behind them" (Nabokov, *King, Queen, Knave* [New York: Putnam, 1968], p. viii). Heidegger suspected that all his previous work would be, or had already been, rendered pointless by the fact that the words he had invented were passing into general use, and thereby being reduced to the level of "concepts," mere tools for accomplishing some purpose extrinsic to them. Analogously, one can imagine Nabokov realizing that his earlier work stood in considerable danger of being classified in the sort of general terms which I am offering here. These are the only sort of terms in which one can do the sort of thing I am trying to do, and which Nabokov despised people for doing – "placing" him in relation to novelists, such as Orwell, who had different gifts and different aims. Nabokov shared Heidegger's hope of eventually coming up with words and books which were so unclassifiable, fell so clearly outside any known way of grouping resemblances and differences, that they would not suffer this sort of banalization. But – as Hegel taught us – no individual achievement of importance escapes such banalization, because "importance" is determined precisely by the degree of effort it takes to bring the particular under the universal, to synthesize the idiosyncratic with the social. The most important achievements are those which make such a synthesis extraordinarily difficult, while nevertheless not making it impossible. Heidegger achieved the perfect balance between initial maximal difficulty of synthesis and eventual transparency, and thereby his acme, in his middle period – the period in which he wrote what he called "the history of Being." Thereafter, in his final period, he became *merely* idiosyncratic, pursuing private crochets, private resonances, private obsessions. Nabokov achieved the same sort of perfect balance in his middle period – the period of *Lolita* and *Pale Fire*. Thereafter, in the period which begins with *Ada* and ends with *Look at the Harlequins!*, he, too, becomes *merely* idiosyncratic. Even in *Ada*, he is talking only to himself half the time. As Robert Alter has said, Ada is a "dazzling, but at times also exasperating, near-masterpiece that lacks the perfect selectivity and control of *Lolita* and *Pale Fire*" ("*Ada*, or the Perils of Paradise," in Peter Quennell, ed., *Vladimir Nabokov: A Tribute* [London: Weidenfeld, 1979], p. 104). The two great novels have a distinctively Nabokovian idiosyncrasy which the earlier novels (except perhaps for *Invitation to a Beheading*) lack, and a perfection of form which the later novels lack.

suspect that only someone who feared that he was executing a partial self-portrait could have made that particular contribution.[27]

Let me offer some further evidence for this interpretation of the two novels by citing another remark from the Afterword to *Lolita*. Nabokov is listing "the nerves of the novel . . . the secret points, the subliminal co-ordinate by means of which the book is plotted" (p. 315). Among these secret points, he tells us, is "the Kasbeam barber (who cost me a month of work)."[28] This barber appears in only one sentence:

> In Kasbeam a very old barber gave me a very mediocre haircut: he babbled of a baseball-playing son of his, and, at every explodent, spat into my neck, and every now and then wiped his glasses on my sheet-wrap, or interrupted his tremulous scissor work to produce faded newspaper clippings, and so inattentive was I that it came as a shock to realize as he pointed to an easelled photograph among the ancient gray lotions, that the moustached young ball player had been dead for the last thirty years. (p. 211)

27 Humbert's nympholepsy and Kinbote's homosexuality are made so plausible and interesting (so "charming," to use the word that everyone in *Bleak House* uses about Skimpole) that – probably just as Nabokov intended – they arouse questions in his readers' minds about Nabokov's own views on sex. I take these to be just more of Nabokov's celebrated false leads. There is certainly something of Nabokov himself in these monsters, but it has nothing in particular to do with any particular sort of sex. Sexual obsessions are just handy examples of a more general phenomenon.

28 The town of Kasbeam is described as seen from a nearby hilltop, in terms which anticipate those used at the climactic moment, just before the end of the novel, when Humbert looks down from another hill to another "toylike" town, the one from which rises the "melody of children at play." Then Humbert realizes that "the hopelessly poignant thing was not Lolita's absence from my side, but the absence of her voice from that concord" (p. 306). This is the moment which produces what Humbert has earlier called the "unbelievable, unbearable, and, I suspect, eternal horror that I know *now*" (p. 167). Humbert, writing his story as he dies of heart disease, describes that horror when he writes: "Alas I was unable to transcend the simple human fact that whatever spiritual solace I might find, whatever lithophanic eternities might be provided for me, nothing could make my Lolita forget the foul lust I had inflicted upon her. Unless it can be proven to me – to me as I am now, today, with my heart and my beard, and my putrefaction – that in the infinite run it does not matter a jot that a North American girl-child named Dolores Haze had been deprived of her childhood by a maniac, unless this can be proven (and if it can, then life is a joke), I see nothing for the treatment of my misery but the melancholy and very local palliative of articulate art. To quote an old poet:

> The moral sense in mortals is the duty
> We have to pay on mortal sense of beauty." (p. 281)

The old poet is Nabokov himself. I am suggesting that he hoped that poets had to pay this duty, but was not sure, and thus not sure that life was not a joke.

I am not sure whether "cost me a month of work" means that Nabokov rewrote the sentence about the barber for a month, or that his associations with the idea of not noticing the death of another's child kept him from writing for a month, or that some actual encounter with another's (perhaps an actual barber's) suffering kept him from writing for a month. It is typical of Nabokov to let his reader guess.

This sentence epitomizes Humbert's lack of curiosity — his inattentiveness to anything irrelevant to his own obsession — and his consequent inability to attain a state of being in which "art," as Nabokov has defined it, is the norm. This failure parallels a failure described earlier in the book, one which occurs when Humbert transcribes from memory the letter in which Charlotte proposes marriage to him, and adds that he has left out at least half of it including "a lyrical passage which I more or less skipped at the time, concerning Lolita's brother who died at two when she was four, and how much I would have liked him" (p. 68).

This is one of only two passages in the book in which Lolita's dead brother is referred to. The other is one in which Humbert complains that Charlotte rarely talks about her daughter — the only subject of interest to him — and in particular that she refers to the dead boy more frequently than to the living girl (p. 80). Humbert mourns that Lolita herself never referred to her pre-Humbertian existence in Humbert's presence. But he did once overhear her talking to a girlfriend, and what she said was: "You know what's so dreadful about dying is that you are completely on your own" (p. 282). This leads Humbert to reflect that "I simply did not know a thing about my darling's mind" and that "quite possibly, behind the awful juvenile clichés, there was in her a garden and a twilight, and a palace gate — dim and adorable regions which happened to be lucidly and absolutely forbidden to me, in my polluted rags and miserable convulsions."

Continuing this meditation on possibilities which had not previously occurred to him, Humbert remembers an occasion on which Lolita may have realized that another of her girlfriends had "such a wonderful fat pink dad and a small chubby brother, and a brand-new baby sister, and a home, and two grinning dogs, and Lolita had nothing" (p. 285). It is left to the reader to make the connection — to put together Lolita's remark about death with the fact that she once had a small, chubby brother who died. This, and the further fact that Humbert does not make the connection himself, is exactly the sort of thing Nabokov expects his ideal readers — the people whom he calls "a lot of little Nabokovs" — to notice. But, ruefully and contemptuously aware that most of his readers will fall short, he tells us in his Afterword what we have missed.

Consider the impact of being told this on the reader who only then remembers that the death of a child is Nabokov's standard example of ultimate pain — the occasion for John Shade's poem "Pale Fire" as well as the central event in *Bend Sinister*. It dawns on this reader that he himself was just as inattentive to that month-long sentence, and to that dead moustached son, as Nabokov suspected he had been. The reader, suddenly revealed to himself as, if not hypocritical, at least cruelly incurious, recognizes his *semblable*, his brother, in Humbert and Kinbote. Suddenly

Lolita does have a "moral in tow."[29] But the moral is not to keep one's hands off little girls but to notice what one is doing, and in particular to notice what people are saying. For it might turn out, it very often does turn out, that people are trying to tell you that they are suffering. Just insofar as one is preoccupied with building up to one's private kind of sexual bliss, like Humbert, or one's private aesthetic bliss, like the reader of *Lolita* who missed that sentence about the barber the first time around, people are likely to suffer still more.

Turning from *Lolita* to *Pale Fire*, we can see Shade as having been given all of Nabokov's own tenderness and kindness and curiosity, but Kinbote as getting all the ecstasy. Shade's poem about the death of his daughter is not nearly as good a poem as *Pale Fire* is a novel. That is because the rest of the novel, Kinbote's commentary, gives us something Shade could not – it surrounds the ordinary suffering of an elderly mortal man with glimpses of Zembla, glimpses of what Humbert Humbert called "a paradise whose skies were the color of hell-flame."[30] Kinbote is a marvel of self-involvement, a man who knows himself to be (except in his dreams) utterly heartless, but who is much more imaginative than Shade. Psychotics *are*, after all, a lot more imaginative than the rest of us. In Humbert and Kinbote, Nabokov managed to create two sociopaths who, unlike most real-life psychotics, managed to write their own case histories, and to do so knowing exactly how those histories would sound to normal ears.

Kinbote is very curious indeed about anything which at all affects his own desires for boys or for glory. He is bored and annoyed by everything else. He is enraged that Shade has dared to write about his own daughter's death and the joy of his own marriage rather than about "the glory of Zembla," about Kinbote's merry minions and his miserable wife. Yet Shade's poem without Kinbote's commentary would be merely wistful. It is the counterpoint between the poem and the commentary which makes the poem itself memorable. Shade's tenderness and kindness are made visible by Kinbote's remorseless pursuit of the sort of

29 Just before giving his definitions of "aesthetic bliss" and of "art" in the Afterword, Nabokov says, "I am neither a reader nor a writer of didactic fiction, and, despite John Ray's assertion, *Lolita* has no moral in tow" (p. 313).

30 In a certain horrible way, Kinbote is absolutely right when he concludes his Foreword to "Pale Fire" by saying, "Without my notes Shade's text simply has no human reality at all since the human reality of such a poem as his (being too skittish and reticent for an autobiographical work) . . . has to depend entirely on the reality of its author and his surroundings, attachments and so forth, a reality that only my notes can provide. To this statement my dear poet would probably not have subscribed, but, for better or worse, it is the commentator who has the last word." Nabokov likes to put truth in the mouths of those who do not realize what they are saying; the Foreword to *Lolita* (by "John Ray, Jr., Ph.D.") is another example of this.

ecstasy which necessarily excludes attention to other people. We are more likely to notice the joys or the sufferings of one person if our attention is directed to it by the surprising indifference of another person. Just as the misery of the peasantry is made visible by the conspicuous consumption of the nobles, or the hovels of the blacks by the swimming pools of the whites, so the death of Shade's daughter is made more vivid by Kinbote's dismissal of it than in Shade's own remembrance. Hegel's point was sound: The thesis will escape our notice, after a bit of time has passed, unless it catches the reflection, the pale fire, of the brand-new, shiny, antithesis.

To put the point in some of Nabokov's favorite terms of praise, Kinbote is, because crueler, cooler and dryer – and thus a better writer – than Shade. Shade's verse, by his own confession, is written above freezing point. In his poem he remarks that his own reputation, among literary critics, is always "one oozy footstep" behind Robert Frost's. Kinbote for once glosses a line with due respect to its author's interests, and speaks for Nabokov when he says, "In the temperature charts of poetry high is low, and low high, so that the degree at which perfect crystallization occurs is above that of tepid facility" (*Pale Fire,* p. 136).[31]

Kinbote understands what Shade is getting at here because, as befits two aspects of a single creator, Shade and Kinbote have a lot in common. Shade realizes this. Cruel as he may be, Kinbote is not vulgar enough to be physically brutal, and to Shade that matters a great deal (p. 145). Shade's knowledge that "without . . . Pride, Lust and Sloth, poetry might never have been born" (p. 150) lets him be indulgent about Kinbote's delusions, as he would not have been indulgent with anyone who

---

31 I am unsure whether Kinbote speaks for Nabokov when he goes on to praise Frost's "Stopping by Woods on a Snowy Evening" and says, "With all his excellent gifts, John Shade could never make *his* snowflakes settle that way." But despite the suspicious terms which Nabokov makes Kinbote use to praise Frost ("a poem that every American boy knows by heart"), I suspect that Nabokov knew perfectly well that he himself could not write poetry as well as Frost could, and consequently that Shade could not either.
  However that may be, Nabokov was very fond of the metaphor of crystallization. Crystal is a different state of being than fluidity, one in which transparency vanishes and is sometimes replaced by iridescence. But the crystals have to be artificial, and as unrepeatable as snowflakes are conventionally supposed to be. Gradus's inability to grasp any but general ideas is paralleled by his inability to like any piece of glass other than homogeneous and transparent ones – such as the –"little hippopotamus made of violet glass" (*Pale Fire,* p. 169) and the "small crystal giraffe" (p. 132) which he prices in the course of his travels. Kinbote nicely describes the form Marxism takes when it becomes a state religion when he says, "Ideas in modern Russia are machine-cut blocks, coming in solid colors; the nuance is outlawed, the interval walled up, the curve grossly stepped." It seems safe to assume that what Gradus admired in the hippopotamus and the giraffe was their lifelikeness – that is, their approximation to the transparent conventional representation of these animals.

brought about physical harm. He treats Kinbote as a fellow artist in whom, as in Swift and Baudelaire, the mind sickened before the body collapsed (p. 111). The two men share the same view of tyrants and fools – of people like Monsieur Pierre, Gradus, and Paduk, whose brutality they take to stem from their underlying vulgarity. This vulgarity consists in being obsessed with general ideas rather than with particular butterflies, words or people.

But although Kinbote is, in a general way, aware of the danger of general ideas, he himself has some very bad ones, whereas Shade really does manage to forswear them all.[32] One of Kinbote's worst ideas is aestheticism, the belief that there is something called "literary technique" or "poetic gift": a practical ability which floats free of the contingencies of an individual poet's life. This is why he thinks that all he need do to gain immortality is to find a good poet, tell the poet all about himself, and then wait to be glorified in imperishable verse. He expects Shade to "merge the glory of Zembla with the glory of his verse" (p. 144), because, as Shade tells him, he thinks that "one can harness words like performing fleas and make them drive other fleas" (p. 144). This idea that somehow language can be separated from authors, that literary technique is a godlike power operating independently of mortal contingencies, and in particular from the author's contingent notion of what goodness is, is the root of "aestheticism" in the bad sense of the term, the

---

32 This forswearing occurs at the passage in "Pale Fire" where Shade gives up his concern for the immortality of the soul, and in particular of his dead daughter's soul, as "flimsy nonsense." Having discovered that the evidence of immortality he thought he had found was based on a misprint, he writes (ll. 806–815):

> But all at once it dawned on me that *this*
> Was the real point, the contrapuntal theme;
> Just this; not text, but texture; not the dream
> But a topsy-turvical coincidence.
> Not flimsy nonsense, but a web of sense.
> Yes! It sufficed that I in life could find
> Some kind of link-and-bobolink, some kind
> Of correlated pattern in the game. . . .

Shade decides that the artist's recognition of contingency, of the absence (or, what comes to the same thing, the utter inscrutability) of any ordering power is preferable to religion's or moral philosophy's claim to have discovered the true name and nature of such a power. By contrast, a taste for general ideas (which Kinbote is unable to realize he shares with Gradus) comes through when Kinbote asks Shade for a "password" and is offered "pity." When Shade refuses to provide a theological backup for this password, Kinbote says, "Now I have caught you, John: once we deny a Higher Intelligence that plans and administrates our individual hereafters we are bound to accept the unspeakably dreadful notion of Chance reaching into eternity" (p. 151). This is precisely the notion Shade *has* accepted in the lines just quoted, a notion whose effects can only be mitigated by what Humbert calls "the very local palliative of articulate art."

sense in which the aesthetic is a matter of form and language rather than of content and life. In this sense of the term, Nabokov the novelist had no interest in being an aesthete, even if Nabokov the theorist could think of no better account of his own practice.

Nabokov has often been read as an aesthete in this sense, and in particular as someone whose work stems from, and illustrates, the weird Barthian view that language works all by itself.[33] Nabokov the theorist and generalizer encourages such a reading, but that reading ignores the point which I take to be illustrated by Nabokov's best practice: Only what is relevant to our sense of what we should do with ourselves, or for others, is aesthetically useful.

One can affirm this point while agreeing with Barthes and his fellow textualists that the point of novels or plays or poems is not to represent human emotions or situations "correctly." Literary art, the nonstandard, nonpredictable use of words, cannot, indeed, be gauged in terms of accuracy of representation. For such accuracy is a matter of conformity to convention, and the point of writing well is precisely to break the crust of convention. But the fact that literary merit is not a matter of reinforcing a widely used final vocabulary, not a matter of success in telling us what we have always known but could not express satisfactorily, should not obscure the fact that literary language is, and always will be, parasitic on ordinary language, and in particular on ordinary moral language. Further, literary interest will always be parasitic on moral interest. In particular, you cannot create a memorable character without thereby making a suggestion about how your reader should act.[34]

I can sum up my reading of Nabokov by saying that he tried to defend himself against the charge of infidelity to his father's project by wielding some general ideas about the function of "the writer," ideas which connect this function both with his own special gifts and with his own special

33 David Rampton and Ellen Pifer both begin their excellent revisionist books on Nabokov by citing, and deploring, a lot of such readings, and by emphasizing the "moral" side of Nabokov. I learned a great deal from both of these books, and in particular from Rampton's discussion of *The Gift*. See Rampton, *Vladimir Nabokov: A Critical Study of the Novels* (Cambridge: Cambridge University Press, 1984), and Pifer, *Nabokov and the Novel* (Cambridge, Mass.: Harvard University Press, 1980).

34 What makes Humbert and Kinbote so much less shadowy and so much more memorable than Cincinnatus or Van Veen is the sheer believability and homeyness of the situations in which they are involved, their interaction with sane people (like Lolita and Shade) rather than simply with their own fantasies or with other fantasts (like Monsieur Pierre or Ada). Cincinnatus is as sympathetic as Shade, and Van as loathsome as Humbert, but in less concrete – and therefore less morally useful – ways. For the concreteness of a character in a novel is a matter of being embedded in situations to which the reader can, out of his own life, imagine analogues.

fear of death. This led him to create a private mythology about a special elite – artists who were good at imagery, who never killed, whose lives were a synthesis of tenderness and ecstasy, who were candidates for literal as well as literary immortality, and who, unlike his father, placed no faith in general ideas about general measures for the general welfare. This was the mythology in which he fruitlessly attempted to enfold Dickens and upon which he relied whenever he was asked, or asked himself, what he had done for the relief of human suffering. But Nabokov also knew perfectly well that his gifts, and artistic gifts generally, neither had any special connection with pity and kindness nor were able to "create worlds."[35] He knew as well as John Shade did that all one can do with such gifts is sort out one's relations to this world – the world in which ugly and ungifted children like Shade's daughter and the boy Jo are humiliated and die. Nabokov's best novels are the ones which exhibit his inability to believe his own general ideas.

35 Nabokov uses this notion of world-creation over and over again. See David Bromwich's "Why Writers Do Not Create Their Own Worlds" (in *Romantic Argument* [Cambridge, Mass.: Harvard University Press, in press]) for an explanation of the drawbacks of this metaphor, one which goes back to Kant and is parasitic on the disastrous Kantian distinction between form and content.

8

The last intellectual in Europe:
Orwell on cruelty

Orwell's last two novels are good examples of what Nabokov thought of as "topical trash," for their importance is a result of having made a big practical difference. We would not now be reading and admiring Orwell's essays, studying his biography, or trying to integrate his vocabulary of moral deliberation into our own unless he had written *Animal Farm* and *1984*. *Lolita* and *Pale Fire* will survive as long as there are gifted, obsessive readers who identify themselves with Humbert and Kinbote. But even Irving Howe, who wrote one of the earliest and best discussions of *1984*, admits that Orwell is one of those writers "who live most significantly for their own age."[1]

Orwell's best novels will be widely read only as long as we describe the politics of the twentieth century as Orwell did. How long that will be will depend on the contingencies of our political future: on what sort of people will be looking back on us, on how events in the next century will reflect back on ours, on how people will decide to describe the Bolshevik Revolution, the Cold War, the brief American hegemony, and the role of countries like Brazil and China. Orwell thought of our century as the period in which "human equality became technically possible" and in which, simultaneously,

. . . practices which had long been abandoned, in some cases for hundreds of years – imprisonment without trial, the use of war prisoners as slaves, public executions, torture to extract confessions, the use of hostages, and the deportation of whole populations – not only became common again, but were tolerated and even defended by people who considered themselves enlightened and progressive.[2]

1 Howe continues: "Such writers, it is possible, will not survive their time, for what makes them so valuable and so endearing to their contemporaries – that mixture of desperate topicality and desperate tenderness – is not likely to be a quality conducive to the greatest art. But it should not matter to us, this possibility that in the future Silone or Orwell will not seem as important as they do for many people in our time. We know what they do for us, and we know that no other writers, including far greater ones, can do it" ("1984: History as Nightmare," in *Twentieth Century Interpretations of 1984*, ed. Samuel Hynes [Englewood Cliffs, N.J.: Prentice-Hall, 1971], p. 53).
2 *The Penguin Complete Novels of George Orwell* (Harmondsworth: Penguin, 1983), p. 861. I shall refer to *1984* in this edition by parenthetical page number. Notice that these

Someday this description of our century may come to seem blinkered or shortsighted. If it does, Orwell will be seen as having inveighed against an evil he did not entirely understand. Our descendants will read him as we read Swift – with admiration for a man who served human liberty, but with little inclination to adopt his classification of political tendencies or his vocabulary of moral and political deliberation. Some present-day leftist critics of Orwell (e.g., Christopher Norris) think that we *already* have a way of seeing Orwell as blinkered and shortsighted. They think that the facts to which he called attention can already be put in a context within which they look quite different. Unlike Norris, I do not think that we have a better alternative context. In the forty years since Orwell wrote, as far as I can see, nobody has come up with a better way of setting out the political alternatives which confront us. Taking his earlier warnings against the greedy and stupid conservatives together with his warnings against the Communist oligarchs, his description of our political situation – of the dangers and options at hand – remains as useful as any we possess.

Nabokov thought aiming at this sort of inevitably temporary utility betrayed the lack, or the waste, of the gifts which were essential to a figure called the "writer." Orwell, too, had views about this mythical figure, pretty much the opposite of Nabokov's views. I urged in Chapter 7 that we set both views aside. Different writers want to do different things. Proust wanted autonomy and beauty; Nietzsche and Heidegger wanted autonomy and sublimity; Nabokov wanted beauty and self-preservation; Orwell wanted to be of use to people who were suffering. They all succeeded. Each of them was brilliantly, *equally*, successful.

Orwell was successful because he wrote exactly the right books at exactly the right time. His description of a particular historical contingency was, it turned out, just what was required to make a difference to the future of liberal politics. He broke the power of what Nabokov enjoyed calling "Bolshevik propaganda" over the minds of liberal intellectuals in England and America. He thereby put us twenty years ahead of our French opposite numbers. They had to wait for *The Gulag Archipelago* before they stopped thinking that liberal hope required the conviction that things behind the Iron Curtain would necessarily get better, and stopped thinking that solidarity against the capitalists required ignor-

practices were common enough outside of Europe – in Africa and Asia, for instance – during the nineteenth century. But Orwell is talking about Europe. As I am also doing in this book, Orwell is being consciously provincial, writing about the particular kinds of people he knows and their moral situation. The provisional title of *1984* was *The Last Man in Europe.*

ng what the Communist oligarchs were doing. Whereas Nabokov sensitized his readers to the permanent possibility of small-scale cruelties produced by the private pursuit of bliss, Orwell sensitized his to a set of excuses for cruelty which had been put into circulation by a particular group – the use of the rhetoric of "human equality" by intellectuals who had allied themselves with a spectacularly successful criminal gang.

The job of sensitizing us to these excuses, of redescribing the post–World War II political situation by redescribing the Soviet Union, was Orwell's great practical contribution. What Howe calls the combination of "desperate tenderness and desperate topicality" in *Animal Farm* and in the first two-thirds of *1984* sufficed to accomplish this limited, practical goal. But in the last third of *1984* we get something different – something not topical, prospective rather than descriptive. After Winston and Julia go to O'Brien's apartment, *1984* becomes a book about O'Brien, not about twentieth-century totalitarian states. This part of the book centers on the citations from *The Theory and Practise of Oligarchical Collectivism* (co-authored by O'Brien) and on O'Brien's explanation of why Winston must be tortured rather than simply shot ("The object of torture is torture"). It is a vision of what Howe calls "post-totalitarianism."[3] It is no longer a warning about what currently is happening in the world, but the creation of a character who illustrates what might someday happen. Orwell was not the first person to suggest that small gangs of criminals might get control of modern states and, thanks to modern technology, stay in control forever. But he was the first to ask how intellectuals in such states might conceive of themselves, once it had become clear that liberal ideals had no relation to a possible human future. O'Brien is his answer to that question.

I want to discuss separately the two jobs Orwell did in his last two novels – redescribing Soviet Russia and inventing O'Brien. I shall begin with the first, returning to O'Brien later. Orwell's admirers often suggest that he accomplished the redescription by reminding us of some plain truths – moral truths whose obviousness is on a par with "two plus two is four." But they are often made nervous by his second accomplishment, and tend, as Howe says, to discount the "apocalyptic desperation" of *1984* and instead to "celebrate [Orwell's] humanity and his 'goodness.' "[4] This goes along with a tendency to suggest that Orwell was not

3 Howe says, "It is extremely important to note that the world of 1984 is *not* totalitarianism as we know it, but totalitarianism after its world triumph. Strictly speaking, the society of Oceania might be called post-totalitarian" (p. 53).
4 "Openly in England, more cautiously in America, there has arisen a desire among intellectuals to belittle Orwell's achievement, often in the guise of celebrating his

really a particularly accomplished writer, but that he made up in good-ness what he lacked in artistry. Here, for example, is Lionel Trilling "Orwell's native gifts are perhaps not of a transcendent kind; they have their roots in a quality of mind that ought to be as frequent as it is modest. This quality may be described as a sort of moral centrality, a directness of relation to moral – and political – fact."[5]

Trilling's way of speaking is echoed by Orwell himself. In a much quoted passage at the end of "Why I Write," Orwell says, "One can write nothing readable unless one constantly strives to efface one's own personality. Good prose is like a window pane."[6] Earlier in the same essay, he lists as one of the four possible motives for writing books the "historical impulse," defined as a "desire to see things as they are, to find out true facts and store them up for the use of posterity" (*CEJL*, I, 4). These passages, and others like them in Orwell's essays, are often read together with the following passage from *1984*:

The Party told you to reject the evidence of your eyes and ears. It was their final, most essential, command. [Winston's] heart sank as he thought of the enormous power arrayed against him, the ease with which any Party intellectual would overthrow him in debate. . . . And yet he was in the right! . . . The obvious, the silly, and the true had got to be defended. Truisms are true, hold on to that! The solid world exists, its laws do not change. Stones are hard, water is wet, objects unsupported fall towards the earth's centre. With the feeling that he was speaking to O'Brien, and also that he was setting forth an important axiom, [Winston] wrote: "Freedom is the freedom to say that two plus two make four. If that is granted, all else follows." (p. 790)

Emphasizing these passages (and others like them)[7] has led many commentators to conclude that Orwell teaches us to set our faces against all those sneaky intellectuals who try to tell us that truth is not "out there," that what counts as a possible truth is a function of the vocabulary you use, and what counts as a truth is a function of the rest of your beliefs. Orwell has, in short, been read as a realist philosopher, a defender of common sense against its cultured, ironist despisers.[8]

humanity and his 'goodness.' They feel embarrassed before the apocalyptic desperation of the book, they begin to wonder whether it may not be just a little overdrawn and humorless, they even suspect it is tinged with the hysteria of the deathbed. Nor can it be denied that all of us would feel more comfortable if the book could be cast out" (p. 42).

5 Trilling, "Orwell on the Future," in *Twentieth Century Interpretations of 1984*, ed. Hynes, p. 24.
6 *The Collected Essays, Journalism and Letters of George Orwell*, I, 7. Hereafter cited parenthetically as *CEJL*.
7 See, for example, *CEJL*, III, 119.
8 Samuel Hynes, for example, sums up the moral of *1984* by saying, "Winston Smith's

On this reading, the crucial opposition in Orwell's thought is the standard metaphysical one between contrived appearance and naked reality. The latter is obscured by bad, untransparent prose and by bad, unnecessarily sophisticated theory. Once the dirt is rubbed off the winowpane, the truth about any moral or political situation will be clear. Only those who have allowed their own personality (and in particular their resentment, sadism, and hunger for power) to cloud their vision will fail to grasp the plain moral facts. One such plain moral fact is that it is better to be kind than to torture. Only such people will try to evade plain epistemological and metaphysical facts through sneaky philosophical maneuvers (e.g., a coherence theory of truth and a holistic philosophy of language – the devices I employed in Chapter 1). Among such facts are that truth is "independent" of human minds and languages, and that gravitation is not "relative" to any human mode of thought.

For reasons already given, I do not think there are any plain moral facts out there in the world, nor any truths independent of language, nor any neutral ground on which to stand and argue that either torture or kindness are preferable to the other. So I want to offer a different reading of Orwell. This is not a matter of wanting to have him on my side of a philosophical argument. He had no more taste for such arguments, or skill at constructing them, than did Nabokov.[9] Rather, it is a matter of insisting that the kind of thing Orwell and Nabokov both did – sensitizing an audience to cases of cruelty and humiliation which they had not noticed – is not usefully thought of as a matter of stripping away appearance and revealing reality. It is better thought of as a redescription of what may happen or has been happening – to be compared, not with reality, but with alternative descriptions of the same events. In the case of the Communist oligarchs, what Orwell and Solzhenitsyn did was to give us an alternative context, an alternative perspective, from which we liberals, the people who think that cruelty is the worst thing we do, could describe the political history of our century.

Deciding between the descriptions which Sartre and Orwell were offering of that history in the late 1940s, like deciding between the descriptions which Fredric Jameson and Irving Howe now offer of our present political situation, is not a matter of confronting or refusing to confront hard, unpleasant facts. Nor is it a matter of being blinded, or

beliefs are as simple as two plus two equal four: the past is fixed, love is private, and the truth is beyond change. All have this in common: they set limits to men's power; they testify to the fact that some things cannot be changed. The point is beyond politics – it is a point of essential humanity" (Hynes, "Introduction" to *Twentieth Century Interpretations of 1984*, ed. Hynes, p. 19).

9 On Orwell's failure to read philosophy, see Bernard Crick, *George Orwell: A Life* (Harmondsworth: Penguin, 1980), pp. 25, 305, 343, 506. See also *CEJL*, III, 98.

not being blinded, by ideology. It is a matter of playing off scenarios against contrasting scenarios, projects against alternative projects, descriptions against redescriptions. Redescriptions which change our minds on political situations are not much like windowpanes. On the contrary, they are the sort of thing which only writers with very special talents, writing at just the right moment in just the right way, are able to bring off. In his better moments, Orwell himself dropped the rhetoric of transparency to plain fact, and recognized that he was doing the same *kind* of thing as his opponents, the apologists for Stalin, were doing. Consider, for example, the following passage:

"Imaginative" writing is as it were a flank attack upon positions that are impregnable from the front. A writer attempting anything that is not coldly "intellectual" can do very little with words in their primary meanings. He gets his effect, if at all, by using words in a tricky roundabout way. (*CEJL*, II, 19)

Orwell's tricky way, in *Animal Farm*, was to throw the incredibly complex and sophisticated character of leftist political discussion into high and absurd relief by retelling the political history of his century in terms suitable for children. The trick worked, because efforts to see an important difference between Stalin and Hitler, and to continue analyzing recent political history with the help of terms like "socialism," "capitalism," and "fascism," had become unwieldy and impracticable. In Kuhnian terms, so many anomalies had been piling up, requiring the addition of so many epicycles, that the overextended structure just needed a sharp kick at the right spot, the right kind of ridicule at the right moment. That was why *Animal Farm* was able to turn liberal opinion around. It was not its relation to reality, but its relation to the most popular alternative description of recent events, that gave it its power. It was a strategically placed lever, not a mirror.

To admirers like Trilling, Orwell provided a fresh glimpse of obvious moral realities. To his Marxist contemporaries, like Isaac Deutscher, and to present-day Marxists like Norris, he was, at best, simpleminded.[10] On my view, Orwell's mind was neither transparent nor simple. It was not *obvious* how to describe the post–World War II political situation, and it still is not. For useful political description is in a vocabulary which sug-

---

10 For an example of the latter reaction, see Isaac Deutscher's discussion of Orwell in "The Mysticism of Cruelty," in *Twentieth Century Interpretations of 1984*, ed. Hynes. For a later use of the "renegade" label, and further doubts about whether Orwell knew enough philosophy, see Norris's "Language, Truth and Ideology: Orwell and the Post-War Left," in *Inside the Myth*, ed. Christopher Norris (London, 1984).

gests answers to the question "What is to be done?" just as useful scientific description is in a vocabulary which increases our ability to predict and control events. Orwell gave us no hints about how to answer Chernyshevsky's question. He merely told us how *not* to try to answer it, what vocabulary to *stop* using. He convinced us that our previous political vocabulary had little relevance to our current political situation, but he did not give us a new one. He sent us back to the drawing board, and we are still there. Nobody has come up with a large framework for relating our large and vague hopes for human equality to the actual distribution of power in the world. The capitalists remain as greedy and shortsighted, and the Communist oligarchs as cynical and corrupt (unless Gorbachev surprises us), as Orwell said they were. No third force has emerged in the world, and neither the neoconservatives nor the post-Marxist left has come up with more than exercises in nostalgia. The possibility that we shall be able to look back on Orwell as blinkered and shortsighted remains, alas, purely theoretical. For nobody has come up with a plausible scenario for actualizing what Orwell called the "technical possibility of human equality."

Such a scenario was what the pre–World War II liberals thought they had. There were times, in the 1930s, when Orwell himself thought he had such a scenario. But the recurrent disconfirmation of his own predictions, his realization that his generation had been suckered by the use of "Marxist theory" as an instrument of Russian politics, and his disgust with cynical prophecies like James Burnham's, led him to write *Animal Farm* and the first two-thirds of *1984*. These books achieved their purpose not by confronting us with moral realities but by making clear to us that we could no longer use our old political ideas, and that we now had none which were of much use for steering events toward liberal goals. All the accusations of "masochistic despair" and "cynical hopelessness" which are flung at Orwell will fall flat until somebody comes up with some new scenarios.

But Orwell did achieve something more than this negative, though necessary and useful, job of sending us back to the drawing board. He did this in the last third of *1984* – the part which is about O'Brien. There he sketched an alternative scenario, one which led in the *wrong* direction. He convinced us that there was a perfectly good chance that the same developments which had made human equality technically possible might make endless slavery possible. He did so by convincing us that nothing in the nature of truth, or man, or history was going to block that scenario, any more than it was going to underwrite the scenario which liberals had been using between the wars. He convinced us that all the intellectual and poetic gifts which had made Greek philosophy, modern science, and Romantic poetry possible might someday find employment in the Ministry of Truth.

In the view of *1984* I am offering, Orwell has no *answer* to O'Brien, and is not interested in giving one. Like Nietzsche, O'Brien regards the whole idea of being "answered," of exchanging ideas, of reasoning together, as a symptom of weakness. Orwell did not invent O'Brien to serve as a dialectical foil, as a modern counterpart to Thrasymachus. He invented him to warn us against him, as one might warn against a typhoon or a rogue elephant. Orwell is not setting up a philosophical position but trying to make a concrete political possibility plausible by answering three questions: "How will the intellectuals of a certain possible future describe themselves?" "What will they do with themselves?" "How will their talents be employed?" He does not view O'Brien as crazy, misguided, seduced by a mistaken theory, or blind to the moral facts. He simply views him as *dangerous* and as *possible*. Orwell's second great achievement, in addition to having made Soviet propaganda look absurd, was to convince the rest of us that O'Brien was, indeed, possible.

As evidence that this way of reading the last part of *1984* is not entirely factitious, I can cite a column which Orwell wrote in 1944. There he dissects what he calls "a very dangerous fallacy, now very widespread in the countries where totalitarianism has not established itself":

The fallacy is to believe that under a dictatorial government you can be free *inside.* . . . The greatest mistake is to imagine that the human being is an autonomous individual. The secret freedom which you can supposedly enjoy under a despotic government is nonsense, because your thoughts are never entirely your own. Philosophers, writers, artists, even scientists, not only need encouragement and an audience, they need constant stimulation from other people. . . . Take away freedom of speech, and the creative faculties dry up. (*CEJL,* III, 133)

How does this passage mesh with the passage from Winston's diary I quoted earlier, the one which concludes, "Freedom is the freedom to say that two plus two equals four. If that is granted, all else follows"? I suggest that the two passages can both be seen as saying that it does not matter whether "two plus two is four" is true, much less whether this truth is "subjective" or "corresponds to external reality." All that matters is that if you do believe it, you can say it without getting hurt. In other words, what matters is your ability to talk to other people about what seems to you true, not what is in fact true. If we take care of freedom, truth can take care of itself. If we are ironic enough about our final vocabularies, and curious enough about everyone else's, we do not have to worry about whether we are in direct contact with moral reality, or whether we are blinded by ideology, or whether we are being weakly "relativistic."

I take Orwell's claim that there is no such thing as *inner* freedom, no such thing as an "autonomous individual," to be the one made by historicist, including Marxist, critics of "liberal individualism." This is that there is nothing deep inside each of us, no common human nature, no built-in human solidarity, to use as a moral reference point.[11] There is nothing to people except what has been socialized into them – their ability to use language, and thereby to exchange beliefs and desires with other people. Orwell reiterated this point when he said, "To abolish class distinctions means abolishing a part of yourself," and when he added that if he himself were to "get outside the class racket" he would "hardly be recognizable as the same person." To be a person is to speak a *particular* language, one which enables us to discuss particular beliefs and desires with particular sorts of people. It is a historical contingency whether we are socialized by Neanderthals, ancient Chinese, Eton, Summerhill, or the Ministry of Truth. Simply by being human we do not have a common bond. For all we share with all other humans is the same thing we share with all other animals – the ability to feel pain.

One way to react to this last point is to say that our moral vocabulary should be extended to cover animals as well as people. A better way, as I suggested in Chapter 4, is to try to isolate something that distinguishes human from animal pain. Here is where O'Brien comes in. O'Brien reminds us that human beings who have been socialized – socialized in any language, any culture – do share a capacity which other animals lack. They can all be given a special kind of pain: They can all be humiliated by the forcible tearing down of the particular structures of language and belief in which they were socialized (or which they pride themselves on having formed for themselves). More specifically, they can be used, and animals cannot, to gratify O'Brien's wish to "tear human minds to pieces and put them together again in new shapes of your own choosing."

The point that sadism aims at humiliation rather than merely at pain in general has been developed in detail by Elaine Scarry in *The Body in Pain: The Making and Unmaking of the World*. It is a consequence of Scarry's argument that the worst thing you can do to somebody is not to make her scream in agony but to use that agony in such a way that even when the agony is over, she cannot reconstitute herself. The idea is to get her to do or say things – and, if possible, believe and desire things, think

11 Or even, I would add, to use as a reference point for clear and distinct ideas about the equality of two two's with four. But this is a philosophical quarrel about the "status" of mathematical truth which need not be pressed here.

thoughts – which later she will be unable to cope with having done or thought. You can thereby, as Scarry puts it, "unmake her world" by making it impossible for her to use language to describe what she has been.

Let me now apply this point to O'Brien's making Winston believe, briefly, that two and two equals five. Notice first that, unlike "Rutherford conspired with the Eurasian General Staff," it is not something O'Brien himself believes. Nor does Winston himself believe it once he is broken and released. It is not, and could not be, Party doctrine. (The book O'Brien co-authored, *The Theory and Practise of Oligarchical Collectivism*, notes that when one is "designing a gun or an airplane" two and two *have* to make four [*1984*, p. 858].) The *only* point in making Winston believe that two and two equals five is to break him. Getting somebody to deny a belief for no reason is a first step toward making her incapable of having a self because she becomes incapable of weaving a coherent web of belief and desire. It makes her irrational, in a quite precise sense: She is unable to give a reason for her belief that fits together with her other beliefs. She becomes irrational not in the sense that she has lost contact with reality but in the sense that she can no longer rationalize – no longer justify herself to herself.

Making Winston briefly believe that two plus two equals five serves the same "breaking" function as making him briefly desire that the rats chew through Julia's face rather than his own. But the latter episode differs from the former in being a final, irreversible unmaking. Winston might be able to include the belief that he had once, under odd conditions, believed that two and two equals five within a coherent story about his character and his life. Temporary irrationality is something around which one can weave a story. But the belief that he once wanted them to *do it to Julia* is not one he can weave a story around. That was why O'Brien saved the rats for the best part, the part in which Winston had to watch himself go to pieces and simultaneously know that he could never pick up those pieces again.

To return to my main point: the fact that two and two does not make five is not the essence of the matter. What matters is that Winston has picked it as symbolic, and that O'Brien knows that. If there were a *truth*, belief in which would break Winston, making him believe that *truth* would be just as good for O'Brien's purposes. Suppose it were the case that Julia had been (like the purported antique dealer, Mr. Charrington) a longtime member of the Thought Police. Suppose she had been instructed by O'Brien to seduce Winston. Suppose that O'Brien told Winston this, giving him no evidence save his own obviously unreliable word. Suppose further that Winston's love for Julia was such that only he same torture which made him able to believe that two and two equals ive could make him believe that Julia had been O'Brien's agent. The ffect would be the same, and the effect is all that matters to O'Brien. Truth and falsity drop out.

O'Brien wants to cause Winston as much pain as possible, and for this purpose what matters is that Winston be forced to realize that he has become incoherent, realize that he is no longer able to use a language or be a self. Although we can say, "I believed something false," nobody can say to himself, "I am, right now, believing something false." So nobody can be humiliated at the moment of believing a falsehood, or by the mere act of having done so. But people can, their torturers hope, experience the ultimate humiliation of saying to themselves, in retrospect, "Now that I have believed or desired *this*, I can never be what I hoped to be, what I thought I was. The story I have been telling myself about myself — my picture of myself as honest, or loyal, or devout — no longer makes sense. I no longer have a self to make sense of. There is no world in which I can picture myself as living, because there is no vocabulary in which I can tell a coherent story about myself." For Winston the sentence he could not utter sincerely and still be able to put himself back together was "Do it to Julia!" and the worst thing in the world happened to be rats. But presumably each of us stands in the same relations to some sentence, and to some thing.

If one can discover that key sentence and that key thing, then, as O'Brien says, one can tear a mind apart and put it together in new shapes of one's own choosing. But it is the sound of the tearing, not the result of the putting together, that is the object of the exercise. It is the breaking that matters. The putting together is just an extra fillip. When Winston comes to love Big Brother, for example, it is irrelevant that Big Brother is in fact unlovable. What matters is that there is no way of going back and forth between a Winston who loves Big Brother and the Winston who loved Julia, cherished the glass paperweight, and could remember the clipping which showed that Rutherford was innocent. The point of breaking Winston is not to bring Winston into line with the Party's ideas. The Inner Party is not torturing Winston because it is afraid of a revolution, or because it is offended by the thought that someone might not love Big Brother. It is torturing Winston for the sake of causing Winston pain, and thereby increasing the pleasure of its members, particularly O'Brien. The only object of O'Brien's intensive seven-year-long study of Winston was to make possible the rich, complicated, delicate, absorbing spectacle of mental pain which Winston would eventually provide. The only point in leaving the thing sitting in the Chestnut Tree Café alive for a while is that it can still feel pain when the telescreen plays "Under the spreading chestnut tree / I sold you and you sold me." Torture is not for the sake of getting people to obey, nor for the sake of getting them to believe falsehoods. As O'Brien says, "The object of torture is torture.'

For a gifted and sensitive intellectual living in a posttotalitarian culture, this sentence is the analogue of "Art for art's sake" or "Truth for its own sake," for torture is now the only art form and the only intellectual discipline available to such a person. That sentence is the central sentence of *1984*. But it is also the one which has been hardest for commentators to handle. Many of them have agreed with John Strachey that

... from the moment when Winston and Julia are, inevitably, caught and their interrogation and torture begins, the book deteriorates. . . . the subject of physical torture, though it was clearly another of his obsessions, was not one with which Orwell was equipped to deal. He had never been tortured, any more than most of the rest of us have been. And those who have no personal experience of this matter may be presumed to know nothing whatever about it.[12]

This last point of Strachey's is, I think, fairly easy to answer. What Strachey neglects is that the last third of *1984* is about O'Brien, not about Winston — about torturing, not about being tortured.

This neglect is the result of a natural desire to identify Orwell with Winston. If we yield to this desire, then passages like the one I quoted earlier, in which Winston insists on the importance of believing that two and two equals four, will be the center of the novel. The last third of the novel will be merely a hysterical and unnecessary tailpiece. The passages I have been emphasizing — the ones in which O'Brien tells about how things look from inside the Inner Party — will be read as reductiones ad absurdum of O'Brien's dialectical position, or else as Raymond Williams reads them. He reads "The object of torture is torture. The object of power is power" as saying that (in a phrase which Orwell had used to

12 John Strachey, "The Strangled Cry," in *Twentieth Century Interpretations of 1984*, ed. Hynes, pp. 58–59. I think that Orwell implicitly answered Strachey when he wrote, "The people who have shown the best understanding of Fascism are either those who have suffered under it or those who have a Fascist streak in themselves" (*CEJL*, II, 172). His biographers have remarked upon Orwell's spurts of sadism. See esp. Crick, *George Orwell*, p. 275n, and also pp. 504, 572. See also Daphne Patai, *The Orwell Mystique: A Study in Male Ideology* (Amherst: University of Massachusetts Press, 1984). Patai argues that sadism was pretty close to the center of Orwell's character; I do not find her case convincing, but she certainly has lots of evidence to cite. Orwell also had a good eye for sadism in others; see his remarks on George Bernard Shaw's sadism at *CEJL*, III, 222. The choice of the name "O'Brien," and the description of O'Brien's physical appearance (*1984*, p. 748), may have been a conscious or unconscious slap at Shaw.

describe James Burnham's position) "power hunger . . . is a natural instinct which does not have to be explained."

Williams recognizes that it is too easy just to identify Orwell with Winston, but he thinks that Orwell's identification with O'Brien was a last-minute self-betrayal. Williams comments:

It is not necessary to deny the existence, even the frequent occurrence, of persecution and power and torture "for their own sake" . . . to go on resisting the cancellation of all links between power and policy. And this cancellation *must* be resisted, if only because it would then be pointless to try to distinguish between social systems, or to inquire, discriminatingly, where this or that system went good or bad.[13]

Williams thinks that if Burnham were right about power hunger's being a natural instinct, there would be no "fact of the matter," no "objective truth" about whether social democracy is better than fascism. He says that Burnham's position "discredits all actual political beliefs and aspirations, since these are inevitably covers for naked power and the wish for it. . . . There is also a cancellation of inquiry and argument, and therefore of the possibility of truth." Williams takes Orwell to have succumbed, briefly and at the last moment, to the pernicious view that there *is* no such possibility. Like Strachey, Williams thinks that the novel goes off the rails at the end.

Quoting Orwell's stricture against Burnham – "Power worship blurs political judgment because it leads, almost unavoidably, to the belief that present trends will continue"– Williams ends his book on Orwell as follows:

Yet Orwell himself, always an opponent of privilege and power, committed himself, in the fiction, to just that submissive belief. The warning that the world could be going that way became, in the very absoluteness of the fiction, an imaginative submission to its inevitability. And then to rattle that chain again is to show little respect to those many men and women, including from the whole record Orwell himself, who have fought and are fighting the destructive and ignorant trends that are still so powerful, and who have kept the strength to imagine, as well as to work for, human dignity, freedom and peace.[14]

Williams's reference to "the strength to imagine . . . human dignity, freedom and peace" brings me back to my claim that we are still at the drawing board. I do not think that we liberals *can* now imagine a future of

---

13 Raymond Williams, *Orwell* (London: Fontana, 1984), pp. 124–125.
14 Ibid., p. 126.

"human dignity, freedom and peace." That is, we cannot tell ourselves a story about how to get from the actual present to such a future. We can picture various socioeconomic setups which would be preferable to the present one. But we have no clear sense of how to get from the actual world to these theoretically possible worlds, and thus no clear idea of what to work for. We have to take as a starting point the world Orwell showed us in 1948: a globe divided into a rich, free, democratic, selfish, and greedy First World; an unchanging Second World run by an impregnable and ruthless Inner Party; and a starving, overpopulated, desperate Third World. We liberals have no plausible large-scale scenario for changing that world so as to realize the "technical possibility of human equality." We have no analogue of the scenario which Nabokov's father, and our grandfathers, had for changing the world of 1900.

This inability to imagine how to get from here to there is a matter neither of loss of moral resolve nor of theoretical superficiality, self-deception, or self-betrayal. It is not something we can remedy by a firmer resolve, or more transparent prose, or better philosophical accounts of man, truth, or history. It is just the way things happen to have fallen out. Sometimes things prove to be just as bad as they first looked. Orwell helped us to formulate a pessimistic description of the political situation which forty years of further experience have only confirmed. This bad news remains the great intransigent fact of contemporary political speculation, the one that blocks all the liberal scenarios.[15]

In contrast to the Strachey-Williams view that the book might have done well to end sooner, I think that the fantasy of endless torture – the suggestion that the future is "a boot stamping on a human face – forever" is essential to *1984,* and that the question about "the possibility of truth" is a red herring. I can outline my own view by taking issue with Williams on three points.

First, I do not think that any large view of the form "political beliefs are really . . ." or "human nature is really . . ." or "truth is really . . ." – any large philosophical claim – *could* discredit political beliefs and aspirations. As I said in Chapter 3, I do not think it is psychologically possible to give up on political liberalism on the basis of a philosophical view about the nature of man or truth or history. Such views are ways of rounding out and becoming self-conscious about one's moral identity, not justifications of that identity or weapons which might destroy it. One

---

15 I think of the European and American left as having tried to evade this fact by taking refuge in theoretical sophistication – acting as if practical scenarios were unnecessary, and as if the intellectuals could fulfill their political responsibilities simply by criticizing obvious evils in terms of ever more "radical" theoretical vocabularies. See my "Thugs and Theorists: A Reply to Bernstein," *Political Theory,* 1987, pp. 564–580.

would have to be very odd to change one's politics because one had become convinced, for example, that a coherence theory of truth was preferable to a correspondence theory. Second, no such view can (in Williams's phrase) "cancel" inquiry, argument, and the quest for truth – any more than it can "cancel" the search for food or for love. Only force can effect such cancellations, not philosophy. Third, one should not read O'Brien as if he were Burnham – a philosopher making large claims about what is "natural." O'Brien is not saying that everything else is a mask for the will to power. He is not saying that the nature of man or power or history insures that that boot will grind down forever, but rather that it just *happens* that it will. He is saying that it just so happens that this is how things came out, and that it just so happens that the scenario can no longer be changed. As a matter of sheer contingent fact – as contingent as a comet or a virus – that is what the future is going to be.

This seems to me the only reading which accords with the fact that O'Brien's account of the future is the part we all remember best about *1984,* the really *scary* part. If we take O'Brien not as making large general claims but as making specific empirical predictions, he is a much more frightening figure. Somewhere we all know that philosophically sophisticated debate about whether human nature is innately benevolent or innately sadistic, or about the internal dialectic of European history, or about human rights, or objective truth, or the representational function of language, is pretty harmless stuff. O'Brien the theorist is about as likely to cause real honest-to-God belly-fear as Burnham or Nietzsche. But O'Brien, the well-informed, well-placed, well-adjusted, intelligent, sensitive, educated member of the Inner Party, is more than just alarming. He is as terrifying a character as we are likely to meet in a book. Orwell managed, by skillful reminders of, and extrapolations from, what happened to real people in real places – things that nowadays we know are still happening – to convince us that O'Brien is a plausible character-type of a possible future society, one in which the intellectuals had accepted the fact that liberal hopes had no chance of realization.

Our initial defense against this suggestion is that O'Brien is a psychologically implausible figure. In this view, the only torturers are insensitive, banal people like Eichmann, Gradus, and Paduk. Anybody who has O'Brien's "curiously civilized" way of settling his spectacles, just *couldn't* have the intentions O'Brien professes. O'Brien is a curious, perceptive intellectual – much like us. Our sort of people don't do that sort of thing.

Orwell showed us how to parry this initial defensive move when he said of H. G. Wells that he was "too sane to understand the modern world." In context, what Orwell meant was that Wells did not have what he called "the Fascist streak" which, he said, Kipling and Jack London had had, and which was necessary to find fascism intelligible.[16] I think that Orwell was half-consciously priding himself on sharing this streak with Kipling. He was priding himself on having the imagination to see that history very well might not go the way he wanted it to go, the way Wells thought it was *bound* to go. But this does not mean that Orwell at any time, even when creating O'Brien, believed that it was *bound* to go that way. The antitheoretical streak in Orwell, which he shared with Nabokov and which made them both unable to take Marxist theory seriously, made him quite certain that things could usually go either way, that the future was up for grabs.

One can see the point of saying that Wells was "too sane" by imagining an optimistic Roman intellectual, living under the Antonines and occupied with charting the progress of humanity from the beginnings of rational thought in Athens to his own enlightened time. He happens to get hold of a copy of the recently collected and edited Christian Scriptures. He is appalled by the psychological implausibility and moral degradation of the figure called "Jesus," for the same reasons that Nietzsche was later to be appalled. When told by an imaginative friend that efforts to emulate this figure may permeate empires larger than Rome's, and may be led by men "who consider themselves enlightened and progressive," he is incredulous. As his friend remarks, he is too sane to grasp the possibility that the world may swerve.[17]

The point of my analogy is that the complex of ideas associated with Christianity – for example, the idea that reciprocal pity is a sufficient basis for political association, the idea that there is something importantly wrong with (to use Orwell's list) "imprisonment without trial, the use of war prisoners as slaves, public executions, torture to extract confessions, the use of hostages, and the deportation of whole populations," the idea that distinctions of wealth, talent, strength, sex, and race are not relevant to public policy – these ideas were once fantasies as implausible as those associated with O'Brien's Oligarchical Collectivism. Once upon a time people like Wilberforce and the Mills would have seemed distasteful hysterical projections of a fantast's morbid imagination. Orwell helps us see that it *just happened* that rule in Europe passed into the hands of people who pitied the humiliated and dreamed of human equality, and

16 See *CEJL*, II, p. 172. Orwell's use of his own sadism to create the character of O'Brien seems to me a triumph of self-knowledge and self-overcoming.

17 In the terminology of Chapter 4, both Wells and my imaginary Roman were metaphysicians – people incapable of seeing their final vocabulary as contingent, and thus driven to believe that something in the nature of reality would preserve that vocabulary.

that it may *just happen* that the world will wind up being ruled by people who lack any such sentiments or ideas. Socialization, to repeat, goes all the way down, and who gets to do the socializing is often a matter of who manages to kill whom first. The triumph of Oligarchical Collectivism, if it comes, will not come because people are basically bad, or really are not brothers, or really have no natural rights, any more than Christianity and political liberalism have triumphed (to the extent they have) because people are basically good, or really are brothers, or really do have natural rights. History may create and empower people like O'Brien as a result of the same kind of accidents that have prevented those people from existing until recently – the same sort of accidents that created and empowered people like J. S. Mill and Orwell himself. That it might be thought importantly wrong to get amusement from watching people being torn apart by animals was once as much an implausible historical contingency as O'Brien's Oligarchical Collectivism. What Orwell helps us see is that it may have *just happened* that Europe began to prize benevolent sentiments and the idea of a common humanity, and that it may *just happen* that the world will wind up being ruled by people who lack any such sentiments and any such moralities.

On my reading, Orwell's denial that there is such a thing as the autonomous individual is part of a larger denial that there is something outside of time or more basic than chance which can be counted on to block, or eventually reverse, such accidental sequences. So I read the passage from Winston's diary about the need to insist that two and two equals four not as Orwell's view about how to keep the O'Briens at bay but, rather, as a description of how to keep ourselves going when things get tight. We do so by talking to other people – trying to get reconfirmation of our own identities by articulating these in the presence of others. We hope that these others will say something to help us keep our web of beliefs and desires coherent. Notice that when Winston wrote in his diary that "everything follows" from the freedom to say that two and two equals four, he had "the feeling that he was speaking to O'Brien." He describes himself as "writing the diary for O'Brien – *to* O'Brien; it was like an interminable letter which no one would ever read, but which was addressed to a particular person and took its color from that fact" (*1984*, p. 79). Notice also that when he is arrested O'Brien tells him that he has "always known" that O'Brien was not on his side, and Winston agrees (p. 880).

Because in an earlier passage Winston says he "knew with more certainty than before, that O'Brien was on his side," this agreement is hard to understand. The best explanation we get of the contradiction comes in a later passage when, just after finally managing to get Winston briefly to believe that two and two equals five, O'Brien asks:

Do you remember writing in your diary that it did not matter whether I was a friend or an enemy, since I was at least a person who understood you and could be talked to? You were right. I enjoy talking to you. Your mind appeals to me. It resembles my own mind except that you happen to be insane. (p. 892)

This passage echoes the first mention of O'Brien in the novel. There we were told that Winston

. . . felt deeply drawn to him, and not solely because he was intrigued by the contrast between O'Brien's urbane manner and his prize-fighter's physique. Much more it was because of a secretly held belief – or perhaps not even a belief, merely a hope – that O'Brien's political orthodoxy was not perfect. Something in his face suggested it irresistibly. And again, perhaps it was not even unorthodoxy that was written in his face, but simply intelligence. (*1984*, p. 748; see also p. 757)

We learn, in the end, that it *was* a hope rather than a belief, and that it *was* intelligence rather than unorthodoxy.

It is tempting to say that this passage, like Winston's abiding and constant love for O'Brien, merely exhibits Winston's masochism, the other side of his sadism.[18] But that would dismiss such love too easily. What the passage does is to remind us that the ironist – the person who has doubts about his own final vocabulary, his own moral identity, and perhaps his own sanity – desperately needs to *talk* to other people, needs this with the same urgency as people need to make love. He needs to do so because only conversation enables him to handle these doubts, to keep himself together, to keep his web of beliefs and desires coherent enough to enable him to act. He has these doubts and these needs because, for one reason or another, socialization did not entirely take. Because his utterances detour through his brain – rather than, as in duckspeak, coming straight from the well-programmed larynx – he has Socratic doubts about the final vocabulary he inherited.[19] So, like Socrates and Proust, he is continually entering into erotic relationships with conversational interlocutors. Sometimes these relationships are masochistic, like Marcel's first relationship to Madame de Guermantes. Sometimes they are sadistic, like Charlus's hoped-for maieutic rela-

---

18 For Winston's sadism, see *1984*, p. 751.
19 For duckspeak, see *1984*, pp. 923, 775. See also the description of Winston in torment at p. 882: "He became simply a mouth that uttered, a hand that signed, whatever was demanded of him"; and compare Scarry, *The Body in Pain*, pp. 49–51.

tionship to Marcel. But which they are is not as important as that these relationships be with people intelligent enough to understand what one is talking about – people who are capable of seeing how one might have these doubts because they know what such doubts are like, people who are themselves given to irony.

This is the function O'Brien serves for Winston. But can one call O'Brien an ironist? Orwell gives him all the standard traits of the British intellectual of Orwell's youth. Indeed, my (unverifiable) hunch is that O'Brien is partially modeled on George Bernard Shaw, an important Socratic figure for Orwell's generation. But unlike Shaw, who shared Nietzsche's taste for the historical sublime, O'Brien has come to terms with the fact that the future will exactly resemble the recent past – not as a matter of metaphysical necessity but because the Party has worked out the techniques necessary to prevent change. O'Brien has mastered doublethink, and is not troubled by doubts about himself or the Party.[20] So he is *not,* in my sense, an ironist. But he still has the *gifts* which, in a time when doublethink had not yet been invented, would have made him an ironist. He does the only possible thing he can with those gifts: He uses them to form the sort of relationship he has with Winston. Presumably Winston is only one of a long series of people, each with a mind like O'Brien's own, whom O'Brien has searched out, studied from afar, and eventually learned enough about to enjoy torturing. With each he has entered into a long, close, intensely felt relationship, in order at the end to feel the pleasure of twisting and breaking the special, hidden, tender parts of a mind with the same gifts as his own – those parts which only he, and perhaps a few of his Minitru colleagues, know how to discover and torment. In this qualified sense, we can think of O'Brien as the last ironist in Europe – someone who is employed in the only way in which the end of liberal hope permits irony to be employed.

I take Orwell to be telling us that whether our future rulers are more like O'Brien or more like J. S. Mill does not depend – as Burnham, Williams, and metaphysicians generally suggest it does – on deep facts about human nature. For, as O'Brien and Humbert Humbert show, intellectual gifts – intelligence, judgment, curiosity, imagination, a taste for beauty – are as malleable as the sexual instinct. They are as capable of

20 I think of doublethink as a kind of deliberately induced schizophrenia, the dwelling of two systems of belief and desire within a single body. One of these systems is able to talk to Winston about his doubts; the other is not. O'Brien shifts back and forth between them in the unconscious way in which those with split personalities can switch into another personality as needed. For further development of this split-personality model of the unconscious, see Donald Davidson's treatment of Freud (discussed in Chapter 2) and my "Freud and Moral Deliberation," in *The Pragmatist's Freud,* ed. Joseph Smith and William Kerrigan.

as many diverse employments as the human hand. The kinks in the brain which provide these gifts have no more connection with some central region of the self – a "natural" self which prefers kindness to torture, or torture to kindness – than do muscular limbs or sensitive genitals. What our future rulers will be like will not be determined by any large necessary truths about human nature and its relation to truth and justice, but by a lot of small contingent facts.

9

Solidarity

If you were a Jew in the period when the trains were running to Auschwitz, your chances of being hidden by your gentile neighbors were greater if you lived in Denmark or Italy than if you lived in Belgium. A
common way of describing this difference is by saying that many Danes
and Italians showed a sense of human solidarity which many Belgians
lacked. Orwell's vision was of a world in which such human solidarity was
– deliberately, through careful planning – made impossible.

The traditional philosophical way of spelling out what we mean by
"human solidarity" is to say that there is something within each of us –
our essential humanity – which resonates to the presence of this same
thing in other human beings. This way of explicating the notion of solidarity coheres with our habit of saying that the audiences in the Coli-
seum, Humbert, Kinbote, O'Brien, the guards at Auschwitz, and the
Belgians who watched the Gestapo drag their Jewish neighbors away
were "inhuman." The idea is that they all lacked some component which
is essential to a full-fledged human being.

Philosophers who deny, as I did in Chapter 2, that there is such a
component, that there is anything like a "core self," are unable to invoke
this latter idea. Our insistence on contingency, and our consequent opposition to ideas like "essence," "nature," and "foundation," makes it
impossible for us to retain the notion that some actions and attitudes are
naturally "inhuman." For this insistence implies that what counts as being
a decent human being is relative to historical circumstance, a matter of
transient consensus about what attitudes are normal and what practices
are just or unjust. Yet at times like that of Auschwitz, when history is in
upheaval and traditional institutions and patterns of behavior are collaps-
ing, we want something which stands beyond history and institutions.
What can there be except human solidarity, our recognition of one another's common humanity?

I have been urging in this book that we try *not* to want something
which stands beyond history and institutions. The fundamental premise
of the book is that a belief can still regulate action, can still be thought
worth dying for, among people who are quite aware that this belief is
caused by nothing deeper than contingent historical circumstance. My

picture of a liberal utopia in Chapter 3 was a sketch of a society in which the charge of "relativism" has lost its force, one in which the notion of "something that stands behind history" has become unintelligible, but in which a sense of human solidarity remains intact. In Chapter 4, my sketch of the liberal ironist was of someone for whom this sense was a matter of imaginative identification with the details of others' lives, rather than a recognition of something antecedently shared. In Chapters 5 and 6, I tried to show how ironist theory can be privatized, and thus prevented from becoming a threat to political liberalism. In Chapters 7 and 8 I tried to show how a loathing for cruelty – a sense that it is the worst thing we do – had been combined, in Nabokov and in Orwell, with a sense of the contingency of selfhood and of history.

In this final chapter I shall say something more general about the claim that we have a moral obligation to feel a sense of solidarity with all other human beings. I start off from a doctrine to which I referred in passing in Chapter 1 – Wilfrid Sellars's analysis of moral obligation in terms of "we-intentions." This analysis takes the basic explanatory notion in this area to be "one of us"[1] – the notion invoked in locutions like "our sort of people" (as opposed to tradesmen and servants), "a comrade in the [radical] movement," a "Greek like ourselves" (as opposed to a barbarian), or a "fellow Catholic" (as opposed to a Protestant, a Jew, or an atheist). I want to deny that "one of us human beings" (as opposed to animals, vegetables, and machines) can have the same sort of force as any of the previous examples. I claim that the force of "us" is, typically, contrastive in the sense that it contrasts with a "they" which is also made up of human beings – the wrong sort of human beings.

Consider, first, those Danes and those Italians. Did they say, about their Jewish neighbors, that they deserved to be saved because they were fellow human beings? Perhaps sometimes they did, but surely they would usually, if queried, have used more parochial terms to explain why they were taking risks to protect a given Jew – for example, that this particular Jew was a fellow Milanese, or a fellow Jutlander, or a fellow member of the same union or profession, or a fellow bocce player, or a

---

1 See Wilfrid Sellars, *Science and Metaphysics*, p. 222: "It is a conceptual fact that people constitute a community, a *we*, by virtue of thinking of each other as *one of us*, and by willing the common good *not* under the species of benevolence – but by willing it as one of us, or from a moral point of view." (For Quinean reasons, I should prefer to bracket "It is a conceptual fact that" in the above quotation, but this metaphilosophical difference from Sellars is irrelevant to the present topic.) Chapter 7 of Sellars's book spells out the implications of this claim. Elsewhere Sellars identifies "we-consciousness" with Christian *caritas* and with Royce's "loyalty." For useful analysis and criticism of Sellars's metaethics, see W. David Solomon, "Ethical Theory," in *The Synoptic Vision: Essays on the Philosophy of Wilfrid Sellars*, ed. C. F. Delaney et al. (Notre Dame, Ind.: University of Notre Dame Press, 1977).

fellow parent of small children. Then consider those Belgians: Surely there were some people whom they *would* have taken risks to protect in similar circumstances, people whom they *did* identify with, under some description or other. But Jews rarely fell under those descriptions. There are, presumably, detailed historicosociological explanations for the relative infrequency among Belgians of fellowship-inspiring descriptions under which Jews could fall — explanations of why "She is a Jewess" so often outweighed "She is, like me, a mother of small children." But "inhumanity," or "hardness of heart," or "lack of a sense of human solidarity" is *not* such an explanation. Such terms, in such a context, are simply shudders of revulsion. Consider, as a final example, the attitude of contemporary American liberals to the unending hopelessness and misery of the lives of the young blacks in American cities. Do we say that these people must be helped because they are our fellow human beings? We may, but it is much more persuasive, morally as well as politically, to describe them as our fellow *Americans* — to insist that it is outrageous that an *American* should live without hope. The point of these examples is that our sense of solidarity is strongest when those with whom solidarity is expressed are thought of as "one of us," where "us" means something smaller and more local than the human race. That is why "because she is a human being" is a weak, unconvincing explanation of a generous action.

From a Christian standpoint this tendency to feel closer to those with whom imaginative identification is easier is deplorable, a temptation to be avoided. It is part of the Christian idea of moral perfection to treat everyone, even the guards at Auschwitz or in the Gulag, as a fellow sinner. For Christians, sanctity is not achieved as long as obligation is felt more strongly to one child of God than to another; invidious contrasts are to be avoided on principle. Secular ethical universalism has taken over this attitude from Christianity. For Kant, it is not because someone is a fellow Milanese or a fellow American that we should feel an obligation toward him or her, but because he or she is a rational being. In his most rigorous mood, Kant tells us that a good action toward another person does not count as a *moral* action, one done for the sake of duty as opposed to one done merely in accordance with duty, unless the person is thought of *simply* as a rational being rather than as a relative, a neighbor, or a fellow citizen. But even if we use neither Christian nor Kantian language, we may feel that there is something morally dubious about a greater concern for a fellow *New Yorker* than for someone facing an equally hopeless and barren life in the slums of Manila or Dakar.

The position put forward in Part I of this book is incompatible with this universalistic attitude, in either its religious or its secular form. It is incompatible with the idea that there is a "natural" cut in the spectrum of similarities and differences which spans the difference between you and a dog, or you and one of Asimov's robots – a cut which marks the end of the rational beings and the beginning of the nonrational ones, the end of moral obligation and the beginning of benevolence. My position entails that feelings of solidarity are necessarily a matter of which similarities and dissimilarities strike us as salient, and that such salience is a function of a historically contingent final vocabulary.

On the other hand, my position is *not* incompatible with urging that we try to extend our sense of "we" to people whom we have previously thought of as "they." This claim, characteristic of liberals – people who are more afraid of being cruel than of anything else – rests on nothing deeper than the historical contingencies to which I referred at the end of Chapter 4. These are the contingencies which brought about the development of the moral and political vocabularies typical of the secularized democratic societies of the West. As this vocabulary has gradually been de-theologized and de-philosophized, "human solidarity" has emerged as a powerful piece of rhetoric. I have no wish to diminish its power, but only to disengage it from what has often been thought of as its "philosophical presuppositions."

The view I am offering says that there is such a thing as moral progress, and that this progress is indeed in the direction of greater human solidarity. But that solidarity is not thought of as recognition of a core self, the human essence, in all human beings. Rather, it is thought of as the ability to see more and more traditional differences (of tribe, religion, race, customs, and the like) as unimportant when compared with similarities with respect to pain and humiliation – the ability to think of people wildly different from ourselves as included in the range of "us." That is why I said, in Chapter 4, that detailed descriptions of particular varieties of pain and humiliation (in, e.g., novels or ethnographies), rather than philosophical or religious treatises, were the modern intellectual's principal contributions to moral progress.

Kant, acting from the best possible motives, sent moral philosophy off in a direction which has made it hard for moral philosophers to see the importance, for moral progress, of such detailed empirical descriptions. Kant wanted to facilitate the sorts of developments which have in fact occurred since his time – the further development of democratic institutions and of a cosmopolitan political consciousness. But he thought that the way to do so was to emphasize not pity for pain and remorse for cruelty but, rather, rationality and obligation – specifically, *moral* obligation. He saw respect for "reason," the common core of humanity, as the only motive which was not "merely empirical" – not dependent on the accidents of attention or of history. By contrasting "rational respect" with feelings of pity and benevolence, he made the latter seem dubious, second-rate motives for not being cruel. He made "morality" something distinct from the ability to notice, and identify with, pain and humiliation.

In recent decades Anglo-American moral philosophers have been turning against Kant. Annette Baier, Cora Diamond, Philippa Foot, Sabina Lovibond, Alasdair MacIntyre, Iris Murdoch, J. B. Schneewind, and others have questioned the basic Kantian assumption that moral deliberation must necessarily take the form of deduction from general, preferably "nonempirical," principles. Most recently, Bernard Williams has tried to distance "morality" – roughly, the complex of notions, centering on that of *obligation,* which we have inherited from Christianity by way of Kant[2] – by calling it a "peculiar institution." It is an institution which refuses to allow that obligations are factors to be weighed along with other ethical considerations in deciding what to do, but instead insists that, as Williams puts it, "only an obligation can beat an obligation."[3] In this view, a moral dilemma can be "rationally" resolved only by finding some higher-order obligation which will outrank lower-ranking, competing obligations – an idea which Schneewind has described as basic to the sort of moral philosophy which looks for what he calls "classical first principles."[4] Williams sums up his attitude toward this peculiar institution in the following passage:

In truth, almost all worthwhile human life lies between the extremes that morality puts before us. It [morality] starkly emphasizes a series of contrasts: between force and reason, persuasion and rational conviction, dislike and disapproval, mere rejection and blame. The attitude that leads it to emphasize all these contrasts can be labeled its *purity.* The purity of morality, its insistence on abstracting the moral consciousness from other kinds of emotional reaction or

2 Williams, in *Ethics and the Limits of Philosophy,* p. 174, says that morality – as the system of ideas centering around a special kind of obligation called "moral" – is "not an invention of the philosophers" but, rather, "the outlook, or, incoherently, part of the outlook, of almost all of us." I take it that by "us" Williams means "people likely to read this book," and in that sense he is quite right in his attribution. On my view, however, this is part of the outlook of most of us in this part of the globe because some of *our* theologians and philosophers invented it.

3 Williams, *Ethics and the Limits of Philosophy,* pp. 180, 187.

4 See J. B. Schneewind, "Moral Knowledge and Moral Principles," in *Knowledge and Necessity,* ed. G. A. Vesey (London and New York: Macmillan, 1970). This essay is reprinted in *Revisions: Changing Perspectives in Moral Philosophy,* ed. Stanley Hauerwas and Alasdair MacIntyre (Notre Dame, Ind.: Notre Dame University Press, 1983), an anthology which contains many good examples of the anti-Kantian tenor of recent Moral philosophy. See especially Iris Murdoch's "Against Dryness" and Annette Baier's "Secular Faith," as well as MacIntyre's introductory essay "Moral Philosophy: What Next?"

social influence, conceals not only the means by which it deals with deviant members of its community, but also the virtues of those means. It is not surprising that it should conceal them, since the virtues can be seen as such only from outside the system, from a point of view that can assign value to it, whereas the morality system is closed in on itself and must consider it an indecent misunderstanding to apply to the system any values other than those of morality itself.[5]

One good example of a view which the "morality system" makes seem indecent is that sketched in Part I of this book: the view that although the idea of a central and universal human component called "reason," a faculty which is the source of our moral obligations, was very useful in creating modern democratic societies, it can now be dispensed with – and *should* be dispensed with, in order to help bring the liberal utopia of Chapter 3 into existence. I have been urging that the democracies are now in a position to throw away some of the ladders used in their own construction. Another central claim of this book, which will seem equally indecent to those who find the purity of morality attractive, is that our responsibilities to others constitute *only* the public side of our lives, a side which competes with our private affections and our private attempts at self-creation, and which has no *automatic* priority over such private motives. Whether it has priority in any given case is a matter for deliberation, a process which will usually not be aided by appeal to "classical first principles." Moral obligation is, in this view, to be thrown in with a lot of other considerations, rather than automatically trump them.

Sellars's view of moral obligations as "we-intentions" gives us a way of firming up both Williams's "moral"-"ethical" distinction and my own public-private distinction. It identifies both with the distinction between ethical considerations which arise from one's sense of solidarity and ethical considerations which arise from, for example, one's attachment to a particular person, or one's idiosyncratic attempt to create oneself anew. For Sellars reconstructs the Kantian obligation-benevolence distinction in a way which avoids the assumption of a central self, the assumption that "reason" is the name for a component present in other human beings, one whose recognition is the explanation of human solidarity.[6]

5 Williams, *Ethics and the Limits of Philosophy*, pp. 194–195.
6 Taking this "essential component of humanity" view at face value has tended to make moral philosophers look like sophistical casuists. This is because we figure out what practices to adopt first, and then expect our philosophers to adjust the definition of "human" or "rational" to suit. For example, we know that we should not kill a fellow human, except in our official capacity as soldier, hangman, abortionist, or the like. So are those whom we *do* kill in those capacities – the armies of the invading tyrant, the serial murderer, the fetus – not human? Well, in a sense, yes and, in a sense, no – but defining the relevant senses is an after-the-fact, largely scholastic exercise. We deliberate about the justice of the war, or the rightness of capital punishment or of abortion,

SOLIDARITY

Instead, he lets us view solidarity as made rather than found, produced in the course of history rather than recognized as an ahistorical fact. He identifies "obligation" with "intersubjective validity" but lets the range of subjects among whom such validity obtains be smaller than the human race. In Sellars's account "intersubjective validity" can refer to validity for all members of the class of Milanese, or of New Yorkers, or of white males, or of ironist intellectuals, or of exploited workers, or of any other Habermasian "communicative community." We can have obligations by virtue of our sense of solidarity with *any* of these groups. For we can have we-intentions, intentions which we express in sentences of the form "*We* all want . . . ," intentions which contrast with those expressed by sentences beginning "*I* want . . . ," by virtue of our membership in any of them, large or small.[7] Sellars's basic idea is that the difference between moral obligation and benevolence is the difference between actual or potential intersubjective agreement among a group of interlocutors and idiosyncratic (individual or group) emotion. Such agreement does not have (*pace* Habermas) any ahistorical conditions of possibility, but is simply a fortunate product of certain historical circumstances.

This is not to say, and Sellars would not say, that the attempt to think in terms of abstractions like "child of God," or "humanity," or "rational being" has done no good. It has done an enormous amount of good, as have notions like "truth for its own sake" and "art for art's sake." Such notions have kept the way open for political and cultural change by providing a fuzzy but inspiring *focus imaginarius* (e.g., *absolute* truth, *pure* art, humanity *as such*). The philosophical problems, and the sense of artificiality associated with these problems, only arise when a handy bit of rhetoric is taken to be a fit subject for "conceptual analysis," when *foci imaginarii* are subjected to close scrutiny – in short, when we start asking about the "nature" of truth, or art, or humanity.

first, and worry later about the "status" of the invader or the murderer or the fetus. When we try to do the opposite, we find that our philosophers offer no sufficient conditions for humanity or rationality less controversial than the original practical questions. It is the *details* of those original questions (just what the invaders have done or will do, just who gets executed and why, just who decides to abort and when) that help us decide what to do. The large general principles wait patiently for the outcome, and then the crucial terms which they contain are redefined to accord with that outcome.

7   Sellars's own interest is not in affirming the fact that "we" may refer to some subset of the class of human or rational beings (e.g., one's tribe) but in preserving the obligation-benevolence distinction within a naturalistic (and, indeed, materialistic) framework, one which makes no reference to a noumenal self, historically unconditioned desires, and so on. I obviously share an interest in the latter attempt, but my main concern here is the former claim. For my present purposes, what is essential is Sellars's idea that "categorical validity" and "moral obligation" can be identified with "being willed as one of us," *independent* of questions about who *we* happen to be.

CRUELTY AND SOLIDARITY

When questions of this sort sound as artificial as they have come to sound since Nietzsche, people may begin to have doubts about the notion of human solidarity. To keep this notion, while granting Nietzsche his point about the contingently historical character of our sense of moral obligation, we need to realize that a *focus imaginarius* is none the worse for being an invention rather than (as Kant thought it) a built-in feature of the human mind. The right way to take the slogan "We have obligations to human beings simply as such" is as a means of reminding ourselves to keep trying to expand our sense of "us" as far as we can. That slogan urges us to extrapolate further in the direction set by certain events in the past – the inclusion among "us" of the family in the next cave, then of the tribe across the river, then of the tribal confederation beyond the mountains, then of the unbelievers beyond the seas (and, perhaps last of all, of the menials who, all this time, have been doing our dirty work). This is a process which we should try to keep going. We should stay on the lookout for marginalized people – people whom we still instinctively think of as "they" rather than "us." We should try to notice our similarities with them. The right way to construe the slogan is as urging us to *create* a more expansive sense of solidarity than we presently have. The wrong way is to think of it as urging us to *recognize* such a solidarity, as something that exists antecedently to our recognition of it. For then we leave ourselves open to the pointlessly skeptical question "Is this solidarity *real?*" We leave ourselves open to Nietzsche's insinuation that the end of religion and metaphysics should mean the end of our attempts not to be cruel.

If one reads that slogan the right way, one will give "we" as concrete and historically specific a sense as possible: It will mean something like "we twentieth-century liberals" or "we heirs to the historical contingencies which have created more and more cosmopolitan, more and more democratic political institutions." If one reads it the wrong way, one will think of our "common humanity" or "natural human rights" as a "philosophical foundation" for democratic politics. The right way of reading these slogans lets one think of philosophy as *in the service* of democratic politics – as a contribution to the attempt to achieve what Rawls calls "reflective equilibrium" between our instinctive reactions to contemporary problems and the general principles on which we have been reared. So understood, philosophy is one of the techniques for reweaving our vocabulary of moral deliberation in order to accommodate new beliefs (e.g., that women and blacks are capable of more than white males had thought, that property is not sacred, that sexual matters are of merely private concern). The wrong way of reading these slogans makes one think of democratic politics as subject to the jurisdiction of a philosophical tribunal – as if philosophers had, or at least should do their best to attain, knowledge of something less dubious than the value of the democratic freedoms and relative social equality which some rich and lucky societies have, quite recently, come to enjoy.

In this book I have tried to work out some of the consequences of the assumption that there can be no such tribunal. On the *public* side of our lives, *nothing* is less dubious than the worth of those freedoms. On the private side of our lives, there may be much which is *equally* hard to doubt, for example, our love or hatred for a particular person, the need to carry out some idiosyncratic project. The existence of these two sides (like the fact that we may belong to several communities and thus have conflicting *moral* obligations, as well as conflicts between moral obligations and private commitments) generates dilemmas. Such dilemmas we shall always have with us, but they are never going to be resolved by appeal to some further, higher set of obligations which a philosophical tribunal might discover and apply. Just as there is nothing which validates a person's or a culture's final vocabulary, there is nothing implicit in that vocabulary which dictates how to reweave it when it is put under strain. All we can do is work with the final vocabulary we have, while keeping our ears open for hints about how it might be expanded or revised.

That is why, at the beginning of Chapter 3, I said that the only argument I could give for the views about language and about selfhood put forward in Chapters 1 and 2 was that these views seemed to cohere better with the institutions of a liberal democracy than the available alternatives do. When the value of such institutions is challenged – not by practical proposals for the erection of alternative institutions but in the name of something more "basic" – no direct answer can be given, because there is no neutral ground. The best one can do with the sort of challenges offered by Nietzsche and Heidegger is make the sort of indirect reply offered in Chapter 5: One can ask these men to *privatize* their projects, their attempts at sublimity – to view them as irrelevant to politics and therefore compatible with the sense of human solidarity which the development of democratic institutions has facilitated. This request for privatization amounts to the request that they resolve an impending dilemma by subordinating sublimity to the desire to avoid cruelty and pain.

In my view, there is nothing to back up such a request, nor need there be. There is no *neutral,* noncircular way to defend the liberal's claim that cruelty is the worst thing we do, any more than there is a neutral way to back up Nietzsche's assertion that this claim expresses a resentful, slavish attitude, or Heidegger's that the idea of the "greatest happiness of the greatness number" is just one more bit of "metaphysics," of the "forgetfulness of Being." We cannot look back behind the processes of socialization which convinced us twentieth-century liberals of the validity of this claim and appeal to something which is more "real" or less ephemeral than the historical contingencies which brought those processes into existence. We have to start from where we are – that is part of the force of Sellars's claim that we are under no obligations other than the "we-intentions" of the communities with which we identify. What takes the curse off this ethnocentrism is not that the largest such group is "humanity" or "all rational beings" – no one, I have been claiming, *can* make *that* identification – but, rather, that it is the ethnocentrism of a "we" ("we liberals") which is dedicated to enlarging itself, to creating an ever larger and more variegated *ethnos*.[8] It is the "we" of the people who have been brought up to distrust ethnocentrism.

To sum up, I want to distinguish human solidarity as the identification with "humanity as such" and as the self-doubt which has gradually, over the last few centuries, been inculcated into inhabitants of the democratic states – doubt about their own sensitivity to the pain and humiliation of others, doubt that present institutional arrangements are adequate to deal with this pain and humiliation, curiosity about possible alternatives. The identification seems to me impossible – a philosopher's invention, an awkward attempt to secularize the idea of becoming one with God. The self-doubt seems to me the characteristic mark of the first epoch in human history in which large numbers of people have become able to separate the question "Do you believe and desire what we believe and desire?" from the question "Are you suffering?" In my jargon, this is the ability to distinguish the question of whether you and I share the same final vocabulary from the question of whether you are in pain. Distinguishing these questions makes it possible to distinguish public from private questions, questions about pain from questions about the point of human life, the domain of the liberal from the domain of the ironist. It thus makes it possible for a single person to be both.

---

8   I develope this point in "Solidarity or Objectivity?" in *Post-Analytic Philosophy*, ed. John Rajchman and Cornel West (New York: Columbia University Press, 1984), and in "On Ethnocentrism: A Reply to Clifford Geertz," *Michigan Quarterly Review* 25 (1986): 525–534.

Index of names

INDEX OF NAMES

1920